Lecture Notes in Mathematics

Edited by A. Dold and B. [illegible]

994

Jean-Lin Journé

Calderón-Zygmund Operators, Pseudo-Differential Operators and the Cauchy Integral of Calderón

Springer-Verlag
Berlin Heidelberg New York Tokyo 1983

Author

Jean-Lin Journé
Ecole Polytechnique, Centre de Mathématiques
91128 Palaiseau Cedex, France

AMS Subject Classifications (1980): 42-02

ISBN 3-540-12313-X Springer-Verlag Berlin Heidelberg New York Tokyo
ISBN 0-387-12313-X Springer-Verlag New York Heidelberg Berlin Tokyo

This work is subject to copyright. All rights are reserved, whether the whole or part of the material is concerned, specifically those of translation, reprinting, re-use of illustrations, broadcasting, reproduction by photocopying machine or similar means, and storage in data banks. Under § 54 of the German Copyright Law where copies are made for other than private use, a fee is payable to "Verwertungsgesellschaft Wort", Munich.

© by Springer-Verlag Berlin Heidelberg 1983
Printed in Germany

Printing and binding: Beltz Offsetdruck, Hemsbach/Bergstr.
2146/3140-543210

INTRODUCTION

In a survey article entitled "Recent progress in classical Fourier analysis", Charles Fefferman wrote in 1974 the following: "Commutator Integrals. Let $D \subseteq C^1$ be a domain bounded by a C^1 curve Γ. Just as in the case of the unit disc, there is a "Hilbert transform" T defined on functions on Γ which sends the real part $u_{|\Gamma}$ of an analytic function $F = u + iv$ to its imaginary part $v_{|\Gamma}$, and it is natural to ask whether T is bounded on $L^2(\Gamma)$ with respect to the arclength measure on Γ. This question is closely connected to the problem of understanding harmonic measure on Γ, i.e., the probability distribution of the place where a particle undergoing Brownian motion starting at a fixed point $P_0 \in D$ first hits Γ.

In effect, T is an integral operator on functions on R^1, given by the formula

$$Tf(x) = \int_{-\infty}^{\infty} \frac{f(y)dy}{(x-y) + i(A(x) - A(y))}$$

with $A \in C^1(R^1)$. Expanding the denominator of the integrand in a geometric series, we obtain T as an infinite sum of operators

$$T_k f(x) = \int_{-\infty}^{\infty} \frac{(A(x) - A(y))^k}{(x-y)^{k+1}} f(y)dy \ .$$

T_k is called the kth commutator integral corresponding to $A(x)$.

Commutator integrals also arise naturally when one tries to construct a calculus of singular integral operators to handle differential equations with nonsmooth coefficients. T_0 is just the Hilbert transform, but already the following two results are deep.

<u>Theorem</u>: Let A be a C^1 function on the line then

(A) (Calderón [14], 1965) T_1 is bounded on L^2.

(B) (Coifman and Y. Meyer 1974, still unpublished) T_2 is bounded on L^2.

To prove (A), Calderón used special contour integration arguments which unfortunately do not apply to higher T_k's. Coifman and Meyer modified and built on Calderón ideas to produce a far more flexible proof, which can probably be pushed further in the near future to cover all the T_k's and possibly T itself."

One can now judge the accuracy of Fefferman's prophecy: see for instance introduction to Chapter 6. Several results connected to these commutators have been obtained recently by A.P. Calderón, R.R. Coifman, G. David, MacIntosh and Y. Meyer. They motivated greatly the topics chosen in a course I taught in Washington University during the academic year 1981-1982.

Indeed, these commutators are the most interesting examples, besides classical convolution operators, of a class of operators introduced by R. Coifman and Y. Meyer in [CM2], under the name "Calderón-Zygmund operators" (CZO's) . Note that their interest lies both in their applications (see [CCFJR], [C3], and [J]) and in the methods one uses to study them, (see [C1], [CM1], and [CMcIM].

Chapters 0 to 3 consist of background material. There, all the tools used in the following chapters are introduced; among which are the Hardy-Littlewood maximal operator, A_p weights of Muckenhoupt, the Calderón-Zygmund decomposition of a function, the spaces H^∞ and BMO, and the Feffermann-Stein #-function.

Chapter 4 is devoted to general properties of CZO's . The essential one is that their L^2-boundedness implies their L^p boundedness for $p \in]1,\infty[$. Substitute results hold for $p=1$ or $p=+\infty$; namely CZO's are bounded from $H^{1,\infty}$ to L^1 and from L^∞ to BMO .

From results of Chapter 4 one can see that the essential problem when dealing with a candidate for a CZO lies in showing the L^2-boundedness. Let us quote the same paper of Charles Feffermann's.

"When neither Plancherel's theorem nor Cotlar's lemma applies, L^2-boundedness of singular operators presents very hard problems, each of which must (so far) be dealt with on its own terms."

That is still true. Chapter 7 illustrates the difficulty of such problems, while in Chapter 8 are explained several specific techniques which already proved to be useful to solve them.

Chapters 5 and 6 treat some connections between the theory of CZO's and the theories of pseudo-differential operators and of Littlewood-Paley.

Since this work is essentially self-contained and not very long, one can easily guess that many applications have not been included. On the other hand, we hope that it is accessible to a wide variety of mathematicians without any specific advanced knowledge in harmonic analysis. Anyhow one should be familiar with the first three chapters of [S] and [SW], for instance, in order to read this set of notes with more ease.

Finally, it is a pleasure to express my thanks to Yves Meyer who introduced me to Harmonic Analysis while teaching a course in Orsay in 78-79, the notes of which I freely used to prepare mine; to Guido Weiss who constantly encouraged and helped me to give a tentatively definite form to this set of notes, and in fact corrected not to say: rewrote-everything (except the present paragraph) and who was, during this whole academic year, the perfect example of the wonderful hospitality of the Washington University Mathematics Department. Thanks also to Richard Rochberg for his irreverent encouragement and occasional help and to Stephen Semmes, who transformed my original unreadible franglish manuscript into a typist's dream. Finally thanks also to Micki Wilderspin who did, as you see, a marvellous typing job.

TABLE OF CONTENTS

CHAPTER 0

PRELIMINARIES ON $L^p_B(R^n, d\mu)$

I. Review of Some Notions of Vector Valued Measurability and Integration.

Let dx denote Lebesgue measure on R^n. We denote by $\omega: R^n \to R$ a weight; that is, a positive, measurable, and locally integrable function. All of the measures on R^n that we shall consider will be of the form $\omega(x)dx$. We shall also write $\omega(E) = \int_E \omega(x)dx$ when $E \subset R^n$ is measurable.

The definition of the spaces $L^p(R^n, \omega dx)$, $1 \leq p \leq \infty$, is well known, and we begin by making precise the definition of these spaces when the functions considered take their values in some Banach space B.

Definition 1: A function $f: R^n \to B$ is simple if it is of the form

$$f = \sum_{i=1}^{N} a_i \chi_{A_i} \quad ,$$

where the a_i's are vectors in B, and the A_i's are bounded measurable subsets of R^n.

We denote the set of simple functions by $S_B(R^n)$.

Definition 2: A function $f: R^n \to B$ is strongly measurable if there exists a sequence (f_n) of simple functions such that

$$\lim_{n \to +\infty} \|f(x) - f_n(x)\|_B = 0 \quad \text{a.e.}$$

Note that if f is strongly measurable and if A is a measurable subset of R^n then $f\chi_A$ is strongly measurable.

Theorem (Pettis [Y]). If f is strongly measurable, then $\|f\|_B : R^n \to R_+$ is measurable.

In view of this result, the spaces $L^p_B(R^n, d\mu)$ can be defined in the following way:

Definition 3: Let $p \in [1,\infty]$. The space $L^p_B(R^n, \omega(x)dx)$ denotes the set of strongly measurable functions f such that $\|f\|_B \in L^p(R^n, \omega(x)dx)$.

Note that $L^p_{loc,B}(R^n, \omega(x)dx)$ can be defined in an analogous way.

Theorem [Y]. For $p \in [1,\infty]$, the space $L^p_B(R^n, \omega(x)dx)$ is a Banach space, its norm being defined by

$$\|f\|_{L^p_B(R^n,\omega(x)dx)} = \| \|f\|_B \|_{L^p(R^n,\omega(x)dx)} .$$

For $p \in [1,\infty[$, $S_B(R^n)$ is a dense subset of $L^p_B(R^n,\omega(x)dx)$.

For all of the function spaces that we shall consider the subscript "0" (e.g. $C^\infty_0(R^n)$) will mean that the functions considered vanish at ∞. Similarly, the subscript "c" (e.g., $L^\infty_{c,B}(R^n)$) will mean that the functions considered vanish in a neighborhood of ∞.

<u>Definition</u> 4: We denote by $C^\infty_{c,B}(R^n)$ the set of functions of the form $f = \sum_{i=1}^N a_i f_i$ where the a_i's are vectors in B and the f_i's are C-valued functions in $C^\infty_c(R^n)$.

As a corollary of the previous theorem, we see that, for $p \in [1,\infty[$, $C^\infty_{c,B}(R^n)$ is also a dense subset of $L^p_B(R^n,\omega(x)dx)$.

A function in $L^1_B(R^n,\omega(x)dx)$ is often called <u>ω-Bochner integrable</u>, and just <u>Bochner integrable</u> if $\omega = 1$.

We denote by $\int_{R^n} \cdot\ \omega(x)dx$ the operator defined from $S_B(R^n)$ to B by

$$\int_{R^n}[\sum_{i=1}^N a_i \chi_{A_i}]\omega(x)dx = \sum_{i=1}^N a_i[\int_{A_i}\omega(x)dx]$$
$$= \sum_{i=1}^N a_i \omega(A_i) .$$

Clearly, this operator extends continuously to $L^1_B(R^n,\omega(x)dx)$ as an operator of norm 1. In particular, if A is a measurable subset of R^n and if $f\chi_A$ if ω-Bochner integrable, then

$$\|\int_A f(x)\omega(x)dx\|_B \leq \int_A \|f(x)\|_B \omega(x)dx .$$

Finally, we have the following well-known result.

<u>Proposition</u> [Y]. <u>Let</u> T <u>be a bounded linear operator from a Banach space</u> B_1 <u>to another such space</u> B_2. <u>For</u> f <u>in</u> $L^1_{B_1}(R^n,\omega(x)dx)$, Tf <u>lies in</u> $L^1_{B_2}(R^n,\omega(x)dx)$, <u>and</u>

$$\int_{R^n} T(f(x))\omega(x)dx = T[\int_{R^n} f(x)\omega(x)dx] .$$

II. <u>The Distribution Function.</u>

Let $f : R^n \to B$ be strongly measurable. Its <u>distribution function</u>, with respect to the measure $\omega(x)dx$, is the function $\alpha_\omega : R_+ \to R_+$ such that

$$\forall \lambda \in R_+ \quad \alpha_\omega(\lambda) = \omega(\{x \in R^n : \|f(x)\|_B > \lambda\}) .$$

All quantities dealing only with the size of f can be expressed in terms of α. In particular,

$$\|f\|^p_{L^p(R^n,\omega dx)} = \int_0^\infty p\lambda^{p-1}\alpha_\omega(\lambda)d\lambda .$$

This simple relation leads to a tool that we shall use several times and which was introduced by Burkholder and Gundy in [BG].

The good λ's inequality. Let $u : R^n \to R^+$ and $v : R^n \to R_+$ satisfy

(i) $\inf(1,u) \in L^p(R^n,\omega(x)dx)$ and $v \in L^p(R^n,\omega(x)dx)$;

(ii) $\exists\, \epsilon > 0$ and $\exists\, \gamma \in [0,1[$ such that $(1+\epsilon)^p\gamma < 1$ and for all $\lambda > 0$,

$$\omega(\{x \in R^n : u(x) > (1+\epsilon)\lambda, v(x) \leq \lambda\})$$
$$\leq \gamma\omega(\{x \in R^n : u(x) > \lambda\}) .$$

Then

$$\|u\|_{L^p(\omega dx)} \leq C(p,\epsilon,\gamma)\|v\|_{L^p(\omega dx)} .$$

<u>Proof</u>: Let us suppose first that we know that $\|u\|_p = \|u\|_{L^p(\omega dx)} < \infty$. Then, using (ii) for the second inequality below,

$$\begin{aligned}\|u\|^p_p &= \int_0^\infty p\lambda^{p-1}\omega(\{u(x)>\lambda\})d\lambda \\ &= (1+\epsilon)^p \int_0^\infty p\lambda^{p-1}\omega(\{u(x) > (1+\epsilon)\lambda\})d\lambda \\ &\leq (1+\epsilon)^p \int_0^\infty p\lambda^{p-1}\omega(\{v(x)>\lambda\})d\lambda \\ &\quad + \gamma(1+\epsilon)^p \int_0^\infty p\lambda^{p-1}\omega(\{u(x)>\lambda\})d\lambda \\ &\leq (1+\epsilon)^p\|v\|^p_p + \gamma(1+\epsilon)^p\|u\|^p_p .\end{aligned}$$

Thus, if we know that $\|u\|^p_p < \infty$, we obtain, since $(1+\epsilon)^p\gamma < 1$,

$$\|u\|^p_p \leq \frac{(1+\epsilon)^p}{1-\gamma(1+\epsilon)^p}\|v\|^p_p .$$

If we do not know that $\|u\|^p_p < \infty$, then we observe that $\inf(1,u)$ satisfies the good λ's inequality and is in $L^p(\mathbb{R}^n, \omega dx)$ by hypothesis. In fact, this is true for $\inf(m,u)$ for any $m > 0$. Hence

$$\|u\|_p^p = \lim_{m\to\infty} \|\inf(m,u)\|_p^p \leqq C(\epsilon,p,v)\|v\|_p^p ,$$

as desired.

Note that it is essential to check the a priori estimate $\inf(1,u) \in L^p(\mathbb{R}^n, \omega dx)$, as is shown by the following example. Take $\epsilon = 1$, ν arbitrarily small and $v = 0$. Then for η small enough, the function $u(x) = \left(\frac{1}{1+|x|}\right)^\eta$ satisfies

$$|\{x : u(x) > 2\lambda\}| \leqq \nu |\{x : u(x) > \lambda\}|$$

for all $\lambda > 0$, but of course $\|u\|_p \neq 0$, and in fact $\|u\|_p = +\infty$. (Here $|A|$ denotes the Lebesgue measure of the set A.) Fortunately, this kind of situation will not occur in the instances where we shall use this inequality.

The distribution function is used in the definition of the Lorentz spaces. (See [SW].) Here we shall only be interested in weak-L^p spaces:

<u>Definition</u> 5: The space weak-$L^p_B(\mathbb{R}^n, \omega(x)dx)$ consists of the strongly measurable functions f for which there exists a constant $C(f)$ such that

$$\omega(\{x \in \mathbb{R}^n : \|f(x)\|_B > \lambda\}) \leqq \frac{(C(f))^p}{\lambda^p}$$

for all $\lambda > 0$.

The smallest $C(f)$ for which the inequality holds is called the weak-L^p "norm" of f, although it is not a true norm. Weak-$L^p_B(\mathbb{R}^n, \omega(x)dx)$ contains $L^p_B(\mathbb{R}^n, \omega(x)dx)$ (strictly) and is included in $L^q_{loc,B}(\mathbb{R}^n, \omega(x)dx)$ for $q < p$.

We shall be dealing primarily with operators defined simultaneously on several L^p spaces, and our interest in these weak-L^p spaces lies mainly in their use in interpolation.

<u>Definition</u> 6: An operator T defined on $L^p_B(\mathbb{R}^n, \omega(x)dx)$ is of weak type (p,p) if it sends $L^p_B(\mathbb{R}^n, \omega(x)dx)$ boundedly into weak-$L^p_B(\mathbb{R}^n, \omega(x)dx)$, that is,

$$\omega(\{x \in \mathbb{R}^n : \|(Tf)(x)\|_B > \lambda\}) \leqq C(T)\, \frac{\|f\|_p^p}{\lambda^p}$$

for some constant $C(T)$ and every $f \in L^p_B(\mathbb{R}^n, \omega(x)dx)$ and every $\lambda > 0$.

When $p = \infty$, we say that T is of weak type (p,p) if it sends L^∞ boundedly into itself. In general, if T takes L^p boundedly into itself, we say that T is of type (p,p). (Thus "weak type" and "type" coincide for $p = \infty$.)

An operator T, defined on some L^p_B space, is said to be sublinear if $|T(\alpha f)| = |\alpha|\,|Tf|$ and $|T(f_1+f_2)| \leq |Tf_1| + |Tf_2|$, for all scalars α and $f, f_1, f_2 \in L^p_B$.

Theorem (Marcinkiewicz interpolation theorem). Let $1 \leq p < q \leq \infty$ and let T be a sublinear operator defined on $L^r_B(R^n, \omega(x)dx)$ for $r = p$ and $r = q$. If T is of weak type (p,p) and (q,q), then it can be defined on L^r_B for $p < r < q$ and is of type (r,r) on this range, and T sends weak $L^r_B(R^n,\omega(x)dx)$ into itself boundedly.

See [SW] for a proof of this theorem and its generalizations.

Finally, let us mention another very simple tool, known as Kolmogorov's inequality, which makes more precise the nature of the inclusion of weak-L^1 into L^p_{loc} for $p < 1$.

Kolmogorov's Inequality:

Let g be in weak-$L^1_B(R^n,\omega(x)dx)$. Then for any measurable subset E of R^n having finite measure and any δ, $0 < \delta < 1$,

$$\int_E \|g\|_B^\delta \omega(x)dx \leq \frac{1}{1-\delta}(\omega(E))^{1-\delta}[C(g)]^\delta .$$

Proof: The desired inequality follows immediately from

$$\int_E \|g\|_B^\delta \omega(x)dx = \int_0^\infty \delta\lambda^{\delta-1}\omega(\{x \in E : \|g(x)\|_B > \lambda\})d\lambda$$

$$\leq \int_0^\infty \delta\lambda^{\delta-1}\inf(\omega(E), \frac{C(g)}{\lambda})d\lambda .$$

III. Rademacher Functions and Extensions of Operators.

Let $t \in [0,1[$ and let $0.\epsilon_1\epsilon_2\epsilon_3,\ldots$ be its dyadic development. (For the t which have more than one such expansion, we choose the development that repeats 0's after some point.) For $m \in N$ we define $r_m(t) = 2\epsilon_m - 1$.

The function r_m is called the mth Rademacher function, and the r_m's are orthogonal in L^2 but they do not form a complete system.

Definition 7: Let B be a Banach space and let $f : [0,1] \to B$ be a strongly measurable function. Then f has a Rademacher expansion if there exists a sequence $(a_i)_{i \in N}$ in $\ell^\infty_B(N)$ such that for almost every $t \in [0,1[$, $\sum_{i=1}^\infty a_i r_i(t)$ converges to $f(t)$.

Note that convergence of the expansion is an important condition on the (a_i), as is shown by the following result.

<u>Theorem</u> (Kahane [K]). If $f : [0,1] \to B$ admits a Rademacher expansion, then for every $p \in [1,\infty[$, f lies in $L^p_B([0,1],dx)$, and there exist $c_p < 1 < C_p$ such that

$$c_p\|f\|_p \leq \|f\|_2 \leq C_p\|f\|_p .$$

Note that c_p and C_p are independent of the Banach space B and of f. When B is a Hilbert space, the orthogonality of the Rademacher system implies that

$$\left(\sum_{i=1}^{\infty} \|a_i\|_B^2\right)^{1/2} = \|f\|_2 .$$

Let $p \in [1,\infty[$, let H_1 and H_2 be two separable Hilbert spaces, and let T be a bounded linear operator from $L^p_{H_1}(R^n,dx)$ to $L^p_{H_2}(R^n,dx)$.

For each sequence $(f_i)_{i\in N}$ of functions in $L^p_{H_1}(R^n,dx)$ we consider the sequence $(Tf_i)_{i\in N}$ in $L^p_{H_2}(R^n,dx)$. This defines an operator that we denote by $T\otimes I$, or by $\tilde{T}$, which is an extension of T to $[L^p_{H_1}(R^n,dx)]^N$. Among these sequences we consider those which belong to $L^p_{\ell^2_{H_1}}(R^n,dx)$; that is, the sequences $(f_i)_{i\in N}$ which satisfy $\|(\sum_{i=1}^{\infty} \|f_i\|^2_{H_1})^{1/2}\|_p < \infty$. Then from the previous theorem we obtain

<u>Corollary</u> 1: $T\otimes I$ <u>is bounded from</u> $L^p_{\ell^2_{H_1}}(R^n,dx)$ <u>to</u> $L^p_{\ell^2_{H_2}}(R^n,dx)$ <u>with a norm no greater than</u> $C(p)\|T\|_{p,p}$, <u>that is</u>

$$\left\|\left[\sum_{i=1}^{\infty} \|Tf_i\|^2_{H_2}\right]^{1/2}\right\|_p \leq C(p)\|T\|_{p,p}\left\|\left[\sum_{i=1}^{\infty} \|f_i\|^2_{H_1}\right]^{1/2}\right\|_p .$$

A particular case of the previous corollary is the following.

<u>Corollary</u> 2: <u>Let</u> (T_i) <u>and</u> (T'_j) <u>be two sequences of bounded linear operators from</u> $L^p(R^n,dx)$ <u>into itself such that for all</u> f <u>in</u> $L^p(R^n)$,

$$\left\|\left[\sum_{i=1}^{\infty} |T_i f|^2\right]^{1/2}\right\|_p \leq C\|f\|_p$$

and

$$\left\|\left[\sum_{i=1}^{\infty} |T'_j f|^2\right]^{1/2}\right\|_p \leq C'\|f\|_p .$$

Then

$$\left\|\left[\sum_{i,j=1}^{\infty} |T_i T'_j f|^2\right]^{1/2}\right\|_p \leq C(p)CC'\|f\|_p .$$

CHAPTER 1

THE HARDY-LITTLEWOOD MAXIMAL OPERATOR

I. The Centered Maximal Function With Respect to Cubes.

Notation: We choose a fixed orthonormal system of co-ordinates on R^n and let Q be any cube on R^n. Then Q is determined by its center and its side length $\delta(Q)$, and we let

$$Q(x,r) = \prod_{1 \leq i \leq n} [x_i - r, x_i + r],$$

so that $\delta(Q(x,r)) = 2r$. For a positive number λ and a cube $Q = Q(x,r)$ we define $\lambda Q = \lambda Q(x,r) = Q(x,\lambda r)$, and for $\lambda = 2$ we write $\bar{Q}$ for $2Q$.

Let ν and μ be two positive Borel measures on R^n. Then $M_\mu(\nu)$ is the function on R^n, taking values in $R_+ \cup \{\infty\}$, defined by

$$M_\mu(\nu)(x) = \sup_{r>0} \frac{\nu(Q(x,r))}{\mu(Q(x,r))} .$$

When μ is Lebesgue measure we denote this function by ν^*, and when ν is of the form $f d\mu$ (where f lies in $L^1_{loc}(R^n, d\mu)$) we often write $M_\mu(f)$ for $M_\mu(\nu)$.

For $f \in L^1_{loc}(R^n, d\mu)$, we let

$$\mu_{Q(x,r)} f = \frac{1}{\mu(Q(x,r))} \int_{Q(x,r)} f \, d\mu .$$

Theorem: The operator M_μ is of type (p,p) for $1 < p \leq \infty$ and of weak type $(1,1)$ (relative to $L^p(R^n, d\mu)$).

This thèorem is of a purely geometric nature; in particular, the estimate for the operator norm of M_μ will depend only on the dimension n and on p.

Because M_μ is clearly of type (∞,∞), it follows from the Marcinkiewicz interpolation theorem (see Section 2 of the preceding chapter) that we only need to prove that M_μ is of weak type $(1,1)$. This result is a special case of the following:

Proposition: If μ and ν are two positive Borel measures, then there exists a constant C depending only on the dimension n such that for all $\lambda > 0$,

$$\mu(\{M_\mu(\nu)(x) > \lambda\}) \leq \frac{C\|\nu\|}{\lambda} :$$

The proof of the proposition is based on the following.

Covering lemma of Besicovitch: (See [GU]) Let A be a bounded subset of R^n, and suppose that for each $x \in A$ we are given a cube Q_x that is centered at x. Then one can extract a possibly finite sequence (Q_k) from the collection $\{Q_x : x \in A\}$ such that

(i) $A \subseteqq \bigcup_k Q_k$

(ii) $\|\Sigma \chi_{Q_k}\|_\infty \leqq C_n$ and

(iii) the collection $\{Q_k\}$ is made up of θ_n subcollections of pairwise-disjoint cubes.

For $\lambda > 0$ fixed, let $A_m = \{x \in R^n : |x| \leqq m$ and $M_\mu(\nu)(x) > \lambda\}$, and, for any $x \in A_m$, let Q_x be any cube centered at x that satisfies $\frac{\nu(Q_x)}{\mu(Q_x)} > \lambda$. By the lemma, there is a sequence (Q_k) of cubes such that

$$\mu(A_m) \leqq \mu(\bigcup Q_k) \leqq \Sigma \mu(Q_k) ,$$

and

$$\begin{aligned} \Sigma \mu(Q_k) &\leqq \frac{1}{\lambda} \Sigma \nu(Q_k) \\ &\leqq \frac{1}{\lambda} \int_{R^n} \Sigma \chi_{Q_k} d\nu \leqq \frac{C_n \|\nu\|}{\lambda} . \end{aligned}$$

The conclusion of the proposition now follows by letting m tend to infinity.

If we denote by $M_{\mu,\nu}$ the operator on $L^1_{loc}(R^n, d\nu)$ defined by

$$M_{\mu,\nu}(f)(x) = \sup_{r>0} \frac{\int_{Q(x,r)} |f| d\nu}{\mu(Q(x,r))} ,$$

then we obtain the following.

Corollary: $M_{\mu,\nu}$ sends $L^1(R^n, d\nu)$ boundedly into weak-$L^1(R^n, d\mu)$.

Notice that if $M_{\mu,\nu}$ took $L^\infty(R^n, d\nu)$ into $L^\infty(R^n, d\mu)$ boundedly, which may not be the case, then the Marcinkiewicz interpolation theorem would apply, despite the change of measure, and we would be able to conclude a result that is similar to the above theorem.

When the measure μ satisfies a certain "homogeneity" property, we can give several different equivalent definitions of the maximal operator.

Definition: A measure μ on R^n is a doubling measure if there exists a constant

C_μ <u>such that for all</u> $x \in R^n$ <u>and all</u> $r > 0$

$$\mu(Q(x,2r)) \leq C_\mu \mu(Q(x,r)) .$$

It is easy to see that this definition remains the same if we replace 2 by any $\alpha > 1$, or if we replace cubes by balls.

The main interest in the case where μ is a doubling measure lies in the following.

Let $\overline{M}_\mu(f)(x) = \sup_{x \in Q} \mu_Q|f|$ for $f \in L^1_{loc}(R^n, d\mu)$. Then we obviously have $\overline{M}_\mu(f)(x) \geq M_\mu(f)(x)$. Furthermore, we also have $\overline{M}_\mu(f)(x) \leq C'_\mu M_\mu(f)(x)$.

To see this, fix x and a cube Q containing x. Then $Q \subseteq Q(x,\delta(Q)) \subseteq 3Q$, so that

$$\mu_Q|f| = \frac{1}{\mu(Q)} \int_Q |f|\, d\mu \leq \frac{1}{\mu(Q)} \int_{Q(x,\delta(Q))} |f|\, d\mu$$

$$\leq \frac{\mu(3Q)}{\mu(Q)} \mu_{Q(x,\delta(Q))}|f|$$

$$\leq C'_\mu M_\mu(f)(x) ,$$

and hence $\overline{M}_\mu(f)(x) \leq C'_\mu M_\mu(f)(x)$.

Observe that the boundedness properties of $\overline{M}_\mu$ now follow immediately from the corresponding properties of M_μ. What makes $\overline{M}_\mu$ easier to work with is the fact that $\{x \in R^n : \overline{M}_\mu(f)(x) > \lambda\}$ is exactly the union of the cubes Q such that $\mu_Q|f| > \lambda$.

As in the above calculation, the doubling measure property of μ is usually used in the following equivalent form: if $N > 0$ and $Q \subseteq Q' \subseteq NQ$ given, then for all $f \in L^1_{loc}(R^n, d\mu)$,

$$\mu_Q|f| \leq C(N,\mu)\mu_{Q'}|f| .$$

Note that $\overline{M}_\mu$ need not have nice boundedness properties if μ is not assumed to be a doubling measure. We shall outline an example below for which the proposition that immediately preceeds the covering lemma of Besicovitch fails. But, first, let us point out that this proposition will still hold for $n = 1$. The proof is the same, except that we use the following covering property that is peculiar to R^1. Let $\{I_\alpha : \alpha \in A\}$ be any collection of intervals on the line. Then there exists a subcollection $\{J_\alpha : \alpha \in A_0\}$ such that $\cup J_\alpha = \cup I_\alpha$ and such that any $x \in R^1$ lies in at most two intervals in the subcollection. (The proof of this fact is easy.)

Hence, in particular, $\overline{M}_\mu$ is of weak type $(1,1)$, and since it is also of type (∞,∞), it is of type (p,p) for $1<p<\infty$ by the Marcinkiewicz interpolation theorem.

For our example, we take $n=2$, $d\mu(x,y) = e^{\sqrt{x^2+y^2}}\,dxdy$, and $\nu = \delta(0)$, the Dirac mass at the origin. For the sake of simplicity, we work with balls instead of cubes. Then the set $\{M_\mu(\nu)>\lambda\}$ is a disc that is centered at the origin which we denote by D_λ. Let D_λ' be the disc with center on the positive y-axis, which passes through the origin, and is tangent to the boundary of D_λ at a point on the positive y-axis. Then one can show that $\mu(D_\lambda') = \frac{1}{\lambda}$ and $\lim_{\lambda\to 0} \frac{\mu(D_\lambda')}{\mu(D_\lambda)} = 0$, so that an estimate of the form $\mu(\{M_\mu(\nu)>\lambda\}) \leqq \frac{C}{\lambda}$ cannot hold.

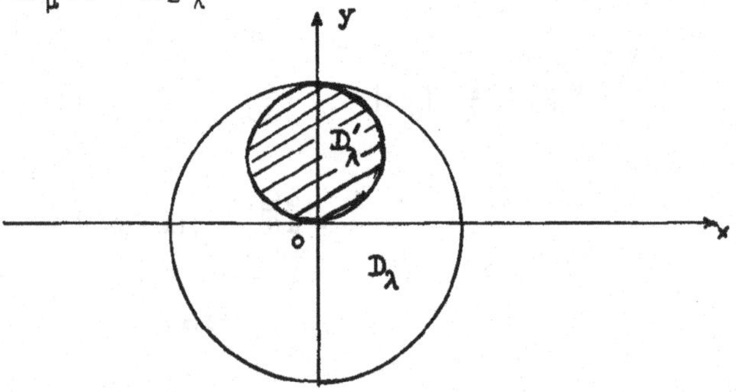

II. <u>The Dyadic Maximal Function.</u>

Another approach to the Hardy-Littlewood maximal operator is through the dyadic maximal function. Again no restriction on the measure is needed, and the classical case can be recaptured if we assume that the appropriate measure is a doubling measure.

Given a cube Q_0 that is "right open", we choose a system of co-ordinates of R^n with respect to which Q_0 is $[0,1[^n$. A cube Q is then a dyadic cube with respect to Q_0 if it is of the form $Q=\{x\in R^n : x=(x_1,\dots x_n)$ with $k_i 2^{-k} \leqq x_i < (k_i+1)2^{-k}\}$, where the k_i's and k range over Z.

We denote by $\Delta(Q_0)$ the collection of all such cubes, and by $\Delta_k(Q_0)$ the collection of all such cubes for a specific $k\in Z$. Then if $Q\in\Delta_k(Q_0)$, we say that it belongs to the kth generation (of Q_0).

The collection of dyadic cubes enjoys some special properties that makes it much easier to work with than the collection of all cubes.

(i) For every $x\in R^n$ and every $k\in Z$, there is a unique $Q\in\Delta_k(Q_0)$ such that $x\in Q$.

(ii) If $Q,Q'\in\Delta(Q_0)$, then either $Q\subseteqq Q'$, $Q'\subseteqq Q$, or $Q\cap Q' = \emptyset$.

(iii) For all $Q\in\Delta_k(Q_0)$ there is a unique $Q'\in\Delta_{k-1}(Q_0)$ such that $Q\subseteqq Q'$.

(We usually denote this Q' by $\tilde{Q}$.) Note that conversely, each cube of the kth generation is the disjoint union of 2^n cubes of the $(k+1)^{th}$ generation.

In particular, it follows from (ii) that the union of a collection of cubes in $\Delta(Q_0)$ which contain a given point is either unbounded or one of the cubes.

Another application of property (ii) which is the most important of the three properties is the following. Let Ω be the union of a collection of cubes in Δ, containing no infinite increasing sequence of cubes. Then Ω can be written as a disjoint union of cubes, each of which is maximal as an element of the given collection of cubes.

It is not true that every cube is included in a dyadic cube; consider, for example, $\bar{Q}_0$ (that is, $2Q_0$). There is, however, a substitute property that usually permits us to reduce a problem about general cubes to a problem about dyadic cubes, when we are working with a doubling measure: for each cube Q there is a dyadic cube Q' such that $Q' \subseteqq Q \subseteqq 5Q'$.

Given a cube Q_0 and a positive measure μ on R^n, we can define a "dyadic" maximal operator $M_{\mu,\delta}$ on $L^1_{loc}(R^n,\mu)$ by

$$M_{\mu,\delta}(f)(x) = \sup_{x \in Q \in \Delta(Q_0)} \mu_Q(|f|) .$$

Note that the set $\{x \in R^n : M_{\mu,\delta}(f)(x) > \lambda\}$ is the union of the dyadic cubes Q for which $\mu_Q|f| > \lambda$.

<u>Theorem</u>: $M_{\mu,\delta}$ <u>is of weak type</u> (1,1) <u>and of type</u> (p,p) <u>for</u> $1 < p \leqq \infty$, <u>(relative to</u> $L^p(R^n,d\mu)$ <u>)</u>.

As before, it is enough to check the weak type estimate. In view of an earlier remark, we may consider the "dyadic decomposition" of the open set $\Omega_\lambda = \{x \in R^n : M_{\mu,\delta}(f)(x) > \lambda\}$. That is, $\Omega_\lambda = \bigcup_{i \in I} Q_i$, where $\{Q_i : i \in I\}$ is a collection of pairwise disjoint dyadic cubes, each of which is maximal as a dyadic cube contained in Ω_λ.

Notice that it is not a priori obvious that $\mu_{Q_i}(|f|) > \lambda$. This is true, however, because Q_i can be written as a disjoint union of cubes Q_i^k that satisfy $\mu_{Q_i^k}|f| > \lambda$.

It now follows easily that $M_{\mu,\delta}$ is of weak type (1,1), since

$$\mu(\Omega_\lambda) = \sum_{i \in I} \mu(Q_i) \leqq \sum_{i \in I} \frac{1}{\lambda} \int_{Q_i} |f|\, d\mu$$

$$\leqq \frac{1}{\lambda} \|f\|_{L^1(R^n, d\mu)} .$$

This proves the theorem, and, in fact, we have obtained a norm estimate for $M_{\mu,\delta}$

that does not depend on μ or on the dimension.

Associated to the dyadic cubes is the well-known Calderón-Zygmund decomposition of a function $f \in L^1_{loc}(\mathbb{R}^n, d\mu)$. We assume that μ is a doubling measure.

Let $\lambda > 0$ and let $(Q_j)_{j \in J}$ be the collection of dyadic cubes contained in $\Omega_\lambda = \{x \in \mathbb{R}^n : M_{\mu,\delta}(f)(x) > \lambda\}$ that are maximal with respect to this inclusion. Let $f_j = (f - \mu_{Q_j} f)\chi_{Q_j}$ and $b = \sum_{j \in J} f_j$. If we put $g = f - b$, we obtain a pair (g,b) of functions which gives us the Calderón-Zygmund decomposition $f = g + b$ corresponding to the value λ and the measure μ.

Lemma 3: (i) $|g| \leqq \lambda$ μ - a.e. on Ω

(ii) For $j \in J$, $\mu_{Q_j}|f| \leqq C(\mu)\lambda$

To prove (ii), we note that $\mu_{Q_j}|f| \leqq C(\mu)\mu_{\tilde{Q}_j}|f|$, since μ is a doubling measure, and that $\mu_{\tilde{Q}_j}|f| \leqq \lambda$, since Q_j is contained in Ω_λ but not $\tilde{Q}_j$. Part (i) is proved using the fact that for any function $f \in L^1_{loc}(\mathbb{R}^n, d\mu)$,

$$M_{\mu,\delta}(f)(x) \geqq |f(x)| \quad \mu\text{ - a.e.}$$

This fact will be proved in the next section.

III. The Connection Between Results on Maximal Functions and Results on Differentiability and Pointwise Convergence.

We begin with an example that makes the connection apparent.

Lebesgue's Differentiation Theorem Let $f \in L^1_{loc}(\mathbb{R}^n, d\mu)$, where μ is a positive Borel measure. Then $\lim_{r \to 0} \mu_{Q(x,r)} f = f(x)$ for μ - a.e. x.

Proof: The idea is to notice first that the above equation holds when f is continuous, and then use the "density" of the continuous functions in $L^1_{loc}(\mathbb{R}^n, d\mu)$ together with the weak type (1,1) inequality.

Let Q be some fixed, but arbitrary cube. Then it suffices to show that for every $\eta > 0$,

$$\mu(Q \cap \{x : \overline{\lim_{r \to 0}} |\mu_{Q(x,r)} f - f(x)| > \eta\}) = 0 .$$

Let $\varepsilon > 0$ be arbitrarily small, and choose a continuous function φ such that

$\|(\varphi - f)\chi_{\bar{Q}}\|_{L^1(\mathbb{R}^n, d\mu)} < \epsilon$.

By writing

$$\mu_{Q(x,r)} f - f(x) = \mu_{Q(x,r)} f - \mu_{Q(x,r)} \varphi$$
$$+ \mu_{Q(x,r)} \varphi - \varphi(x)$$
$$+ \varphi(x) - f(x) \quad ,$$

we see that $Q \cap \{x : \overline{\lim_{r\to 0}} |\mu_{Q(x,r)} f - f(x)| > \eta\}$, whose μ-measure we must estimate, is within the union of three sets E_1, E_2, and E_3 , where

$$E_1 = Q \cap \{x : \overline{\lim_{r\to 0}} |\mu_{Q(x,r)} f - \mu_{Q(x,r)} \varphi| > \tfrac{\eta}{3}\} \quad ,$$

$$E_2 = Q \cap \{x : \overline{\lim_{r\to 0}} |\mu_{Q(x,r)} \varphi - \varphi(x)| > \tfrac{\eta}{3}\} \quad ,$$

and

$$E_3 = Q \cap \{x : |f(x) - \varphi(x)| > \tfrac{\eta}{3}\} \quad .$$

We have $\mu(E_1) \leqq \frac{3}{\eta} C_\mu \|(f - \varphi)\chi_{\bar{Q}}\|_1 \leqq \frac{3\epsilon}{\eta} C_\mu$ (by the weak type (1,1) estimate for M_μ ; see the theorem in section I), $\mu(E_2) = 0$, and $\mu(E_3) \leqq \frac{3}{\eta}\|(f - \varphi)\chi_{\bar{Q}}\|_1 \leqq \frac{3\epsilon}{\eta}$. Hence, because $\epsilon > 0$ is arbitrary, $\mu(E_1 \cup E_2 \cup E_3)$ can be made arbitrarily small; this proves the theorem.

In particular, $M_\mu f(x) \geqq |f(x)|$ μ - a.e.

If we assume that μ is a doubling measure, then we may replace cubes in the above proof by balls that are not necessarily centered at x . The proof is essentially the same, except that we use the weak type (1,1) estimate for the noncentered maximal function. Similarly, we could prove a version of this theorem for dyadic cubes using the results of the preceeding section for the dyadic maximal function. From this we may conclude that $\bar{M}_{\mu,\delta} f(x) \geqq |f(x)|$ μ - a.e. , a fact needed at the end of the last section.

The pattern of the above proof may be reproduced to obtain other differentiability results from estimates on maximal functions. This motivates the search for extensions of the original maximal theorem of Hardy and Littlewood. Here are some simple examples.

For the moment, the measure on $\mathbb{R}^n$ under consideration is Lebesgue measure.

<u>Proposition</u>: <u>Suppose that</u> $\varphi : \mathbb{R}^n \to \mathbb{C}$ <u>is dominated by a decreasing radial function that is integrable.</u> <u>For</u> $\epsilon > 0$ <u>let</u> $\varphi_\epsilon(x) = \frac{1}{\epsilon^n}\varphi(\frac{x}{\epsilon})$. Then, for $1 < p < \infty$,

$$\| \sup_{\epsilon>0} |\varphi_\epsilon * f(x)| \|_p \leq C(p,\varphi)\|f\|_p \quad .$$

See [SW] for a proof, which is effected by showing that $\sup_{\epsilon>0} |\varphi_\epsilon * f|$ is dominated by f^* pointwise. Hence a weak type $(1,1)$ result holds as well.

We denote by $\mathcal{R}$ the family of all rectangles with sides parallel to the co-ordinate axes, that is, Cartesian products of intervals.

Let $f^{**}(x) = \sup_{\substack{R\in\mathcal{R}\\ x\in R}} m_R|f|$. The operator $f \to f^{**}$ is called the <u>strong maximal operator</u>.

<u>Theorem</u>: <u>The strong maximal operator is of type</u> (p,p) <u>for</u> $p>1$.

The strong maximal operator, however, is not of weak type $(1,1)$.

This theorem is proved by "iterating" the one-dimensional result. Define $m_1,\dots,m_n$ on $L^1_{loc}(R^n,dx)$ by setting

$$m_i f(x_1,\dots,x_n)$$

$$= \sup_{\substack{h>0\\ k>0}} \frac{1}{h+k}\int_{x-k}^{x+h} |f(x_1,\dots,x_{i-1},y,x_{i+1},\dots,x_n)|\,dy \ .$$

From our one-dimensional results it follows that each m_i is bounded on $L^p(R^n,dx)$, $1<p\leq\infty$, and it is easy to show that

$$f^{**}(x) \leq m_1(m_2(m_2(\dots(m_n f))\dots)(x) \quad ,$$

which implies the theorem.

Later on we shall see another application of these "partial maximal operators" (i.e., the $m_1,\dots,m_n$) .

To this strong maximal operator corresponds a convolution operator of the type $\sup_{\epsilon>0} |\varphi_\epsilon * f|$, but here the co-ordinates have to be considered independently:

<u>Proposition</u>: <u>Let</u> $\psi_1,\dots,\psi_n : R_+ \to R_+$ <u>be</u> n <u>integrable decreasing functions, and let</u> $\varphi : R^n \to C$ <u>satisfy</u>

$$|\varphi(x_1,\dots,x_n)| \leq \prod_{i=1}^{n} \psi_i(|x_i|) \quad .$$

<u>For positive numbers</u> $\epsilon_1,\dots,\epsilon_n$ <u>we define</u> $\varphi_{\epsilon_1,\dots,\epsilon_n}$ <u>by</u>

$$\varphi_{(\epsilon_i)}(x_1,\dots,x_n) = \varphi_{\epsilon_1,\dots,\epsilon_n}(x_1,\dots,x_n) = \frac{1}{\prod_{j=1}^{n}\epsilon_i}\,\varphi\left(\frac{x_1}{\epsilon_1},\frac{x_2}{\epsilon_1},\dots,\frac{x_n}{\epsilon_n}\right).$$

Then, for $1<p<\infty$,

$$\|\sup_{\epsilon_i>0} |\varphi_{(\epsilon_i)}*f|\,\|_p \leqq C\|f\|_p \ .$$

This result is proved essentially by repeating the proof of the analogous earlier proposition for $\sup_{\epsilon>0}|\varphi_\epsilon *f|$. One proves that this new maximal operator is pointwise dominated by the strong maximal operator.

The arguments used in the proofs of these three results are well known, however, the proof of the following result requires some more original techniques:

Theorem: Let (θ_j) be a lacunary sequence tending to zero (i.e. $\overline{\lim}\,\frac{\theta_{j+1}}{\theta_j}<1$). Let $a_1,\dots,a_n,\lambda_1,\dots,\lambda_n$ be $2n$ positive real numbers, and for each $t>0$ define $\gamma(t)\in\mathbb{R}^n$ by

$$\gamma(t) = (\lambda_1 t^{a_1},\dots,\lambda_n t^{a_n}) \ .$$

For $f\in L^1_{loc}(\mathbb{R}^n,dx)$ let Mf denote the maximal function defined by

$$Mf(x) = \sup_{\substack{h>0\\ j\in N}} \int_{-h}^{h} |f(x-t\gamma(\theta_j))|\,\frac{dt}{h} \ .$$

Then M is of type (p,p) for $p>1$.

See [NSW] for a proof of the theorem as well as some consequences. Note that M is a supremum of partial maximal operator, and the fact that $(\theta_j)_{j\in N}$ is lacunary is crucial.

CHAPTER 2

A_p WEIGHTS

I. PAIRS OF WEIGHTS.

The operators that we shall study in the next chapters will often be controlled in some way by the Hardy-Littlewood maximal function with respect to Lebesgue measure. Because of this fact, it is worthwhile to study further the boundedness properties of the maximal function. (Although we shall be working with Lebesgue measure, we could just as easily work with some doubling measure, making only obvious changes when necessary.)

Let us begin with the following question (See [M]): Suppose that $1 \leqq p < +\infty$. For which pairs of weights (u,v) on R^n is the Hardy-Littlewood maximal operator bounded from $L^p(R^n, vdx)$ into weak-$L^p(R^n, udx)$? That is, for which such pairs (u,v) does there exist a $C > 0$ such that for every $\lambda > 0$ and every $f \in L^1_{loc}(R^n, dx) \cap L^p(R^n, vdx)$ it is true that

$$(1) \qquad \int_{\{x : f^*(x) > \lambda\}} udx \leqq \frac{C(u,v)}{\lambda^p} \int_{R^n} |f|^p vdx \ ?$$

Suppose first that $1 < p < \infty$. Then there is a very simple necessary and sufficient condition on the pair (u,v) : $\exists\ C > 0$ such that for every cube Q,

$$(2) \qquad m_Q u [m_Q v^{-\frac{1}{p-1}}]^{p-1} \leqq C .$$

To see that (1) implies (2), let Q be a cube and let $\epsilon > 0$ be arbitrary, so that $(v+\epsilon)^{-\frac{1}{p-1}} \in L^p_{loc}(R^n, dx)$. Then, in (1), we take $f = (v+\epsilon)^{-\frac{1}{p-1}} \chi_Q$ and $\lambda = m_Q(v+\epsilon)^{-\frac{1}{p-1}}$, so that $Q \subseteq \{x : f^*(x) > \lambda\}$, and, hence,

$$\int_Q udx \leqq \frac{C}{[m_Q(v+\epsilon)^{-\frac{1}{p-1}}]^p} \int_Q (v+\epsilon)^{1-\frac{1}{p-1}} dx$$

Thus, $m_Q(u) \leqq C[m_Q(v+\epsilon)^{-\frac{1}{p-1}}]^{1-p}$, or $m_Q(u)[m_Q(v+\epsilon)^{-\frac{1}{p-1}}]^{p-1} \leqq C$.

Because this is independent of $\epsilon > 0$, we obtain (2).

Conversely, let us assume now that (2) holds. For the moment we take f^* to be the centered maximal function; we loose no generality by doing this, since the reference measure is Lebesgue measure (a doubling measure).

For $f \in L^1_{loc}(\mathbb{R}^n, vdx)$, we put

$$M_{u,v}f(x) = \sup_{r>0} \frac{\int_{Q(x,r)} f\, vdy}{\int_{Q(x,r)} udy} .$$

Because we know that $M_{u,v}$ sends $L^1(\mathbb{R}^n, vdx)$ into weak-$L^1(\mathbb{R}^n, udx)$ boundedly (see page 20) the condition (1) will follow from the inequality

$$(3) \qquad f^* \leqq C[M_{u,v}|f|^p]^{\frac{1}{p}} ,$$

which we shall now prove.

Let Q be a cube. We write $f = (f\, v^{1/p})\, v^{-\frac{1}{p}}$ and apply Hölder's inequality to obtain that

$$\frac{1}{|Q|}\int_Q |f|dx \leqq [\frac{1}{|Q|}\int_Q |f|^p vdx]^{\frac{1}{p}} [\frac{1}{|Q|}\int_Q v^{-\frac{1}{p-1}} dx]^{\frac{p-1}{p}} .$$

Applying (2), we thus obtain

$$m_Q|f| \leqq C \frac{[m_Q|f|^p v]^{\frac{1}{p}}}{[m_Q u]^{\frac{1}{p}}}$$

and hence (3) follows by taking the supremum over all Q centered at x .

Although we shall be most interested in the case $u = v$, the case $u \neq v$ has interesting applications. Among them is a proof of Rubio de Francia's extrapolation theorem, which we shall state on page 32. (See [R] and [G]).

Let us now consider $p = 1$. The "limit" of (2) as p tends to 1 gives the following condition:

There exists $C > 0$ such that for all cubes Q ,

$$(2)' \qquad m_Q(u)(\operatorname*{ess\,inf}_{x \in Q} v(x))^{-1} \leqq C .$$

This turns out to be the correct condition equivalent to (1) with $p = 1$, and it may be reformulated more simply as $u^* \leqq C v$ a.e.

To see this assume first that (1) holds for $p = 1$. Let $Q_1 \subsetneqq Q_2$ and $f = \chi_{Q_1}$.

Then $y \in Q_2$ implies that $f^*(y) \geqq \frac{|Q_1|}{|Q_2|}$, so that

$$\int_{Q_2} u\,dx \leqq \int_{f^* \geqq \frac{|Q_1|}{|Q_2|}} u\,dx$$

$$\leqq \frac{C}{\left(\frac{|Q_1|}{|Q_2|}\right)} \int_{R^n} \chi_{Q_1} v\,dx ,$$

and, hence,

$$\frac{1}{|Q_2|} \int_{Q_2} u\,dx \leqq \frac{C}{|Q_1|} \int_{Q_1} v\,dx .$$

Now, using the fact that $\lim_{Q_1 \to \{x\}} m_{Q_1}(v) = v(x)$ almost everywhere in Q_2, we obtain

$$m_{Q_2}(u) \leqq C \operatorname*{ess\,inf}_{x \in Q_2} v(x) ,$$

as desired.

Conversely, suppose now that (2)′ holds. Then, as before, we take f^* to be the centered maximal function, and it is enough to prove that

$$f^* \leqq C\, M_{u,v} f .$$

But

$$\frac{1}{|Q|} \int_Q |f|\,dx \leqq \left(\frac{1}{|Q|} \int_Q |f|\,dx\right)\left(\frac{C \operatorname*{ess\,inf}_{y \in Q} v(y)}{m_Q(u)}\right)$$

$$\leqq C \frac{\frac{1}{|Q|} \int_Q |f| v\,dx}{\frac{1}{|Q|} \int_Q u\,dx} ,$$

and the inequality that we wanted now follows from taking the supremum over Q centered at x on both sides of this last inequality.

II. THE CASE $u = v$ A_p WEIGHTS.

Definition. A weight ω is said to be in A_1 if there exists $C > 0$ such that $\omega^*(x) \leq C\omega(x)$ almost everywhere, or, equivalently, for each cube Q, $m_Q(\omega) \leq C \operatorname{ess\,inf}_{x\in Q} \omega(x)$. The constant C is sometimes written $C_1(\omega)$.

Hence $\omega \in A_1$ if and only if $u = v = \omega$ satisfies (2)′ above.

Theorem (Coifman-Rochberg [CR]).

(i) If $0 \leq \delta < 1$ and $f \in L^1_{loc}(R^n, dx)$, then $(f^*)^\delta \in A_1$ with $C_1((f^*)^\delta) \leq C_\delta$ for some constant C_δ (depending only on δ).

(ii) If $\omega \in A_1$, then there exist $0 \leq \delta < 1$ and $C > 0$, and there exists $f \in L^1_{loc}(R^n, dx)$ and $h > 0$ such that $\|h\|_\infty \leq C$, $\|h^{-1}\|_\infty \leq C$, and

$$\omega = h(f^*)^\delta .$$

We shall prove part (ii) on p. 32 (near the end of this section), after we have developed the theory of weights more fully. As for (i), suppose that $0 \leq \delta < 1$ and $f \in L^1_{loc}(R, dx)$ have been given. For any fixed cube Q we set $f_1 = f\chi_{\overline{Q}}$ and $f_2 = f - f_1$. (Recall that $\overline{Q}$ denotes the double of the cube Q.) Because $(f^*)^\delta \leq 2^\delta[(f_1^*)^\delta + (f_2^*)^\delta]$, we need only prove that

$$m_Q(f_i^*)^\delta \leq C \operatorname*{ess\,inf}_{x\in Q} (f^*)^\delta(x)$$

for $i = 1,2$.

For $i = 1$ we use Kolmogorov's inequality and the fact that f_1^* is in weak-$L^1(R^n, dx)$, with norm not exceeding $C\|f_1\|_1$. Hence,

$$\int_Q (f_1^*)^\delta dx \leq C_\delta\, |Q|^{1-\delta} \|f_1\|_1^\delta ,$$

and, thus,

$$m_Q((f_1^*)^\delta) \leq C_\delta\left[\frac{\|f_1\|_1}{|Q|}\right]^\delta \leq C_\delta \inf_{x\in Q} (f^*)^\delta .$$

Suppose now that $i = 2$. Observe that for $x \in Q$, $f_2^*(x) = \sup\{m_{Q'}|f_2| : x \in Q'$ and $Q' \cap \overline{Q}^c \neq \emptyset\}$, and that for such Q', $Q \subseteq 5Q'$.

Fix $x_0 \in Q$ and let x be any point in Q. Then

$$f_2^*(x) = \sup_{Q' \ni x} m_{Q'}|f_2| \leq C \sup_{Q' \ni x} m_{5Q'}|f_2| \leq C f_2^*(x_0) .$$

Since x_0 was fixed but arbitrary, we conclude that

$$f_2^*(x) \leq C \inf_{x_0 \in Q} f_2^*(x_0) ,$$

and hence

$$m_Q(f_2^*)^\delta \leq C \inf_{x_0 \in Q} (f^*(x_0))^\delta ,$$

which is what we wanted.

As an application of this theorem, we consider the following result (a general version can be found in [R]):

<u>Proposition</u>: <u>Let</u> T <u>be a linear operator with the property that for every</u> $C > 0$ <u>there is a</u> $C' > 0$ <u>such that</u> $\omega \in A_1$ <u>and</u> $C_1(\omega) \leq C$ <u>imply that</u>

$$\|T\|_{L^p(R^n,\omega dx), L^p(R^n,\omega dx)} \leq C' .$$

<u>Then</u> T <u>is bounded on all</u> $L^q(R^n, dx)$ <u>for</u> $\infty > q > p$.

We are not going to prove this result in detail, but, rather, we shall content ourselves with giving the idea of the proof. In view of this we shall assume that T is bounded on $L^q(R^n, dx)$, and use this fact to obtain an estimate on the operator norm of T on $L^q(R^n, dx)$ that depends only on p, q, C, and C'. (Making such an a priori assumption is a common practice but we shall not work to get rid of this assumption, since we are only interested in giving an idea of how the proof works.)

So let $f \in C_0^\infty(R^n)$, $p < q < \infty$, and suppose that $s > 1$. (We shall choose s later; one should think of it as being close to 1 .) Then,

$$\int_{R^n} |Tf|^q dx = \int_{R^n} |Tf|^p [|Tf|^{(q-p)s}]^{\frac{1}{s}} dx$$

$$\leq \int_{R^n} |Tf|^p [(|Tf|^{(q-p)s})^*]^{\frac{1}{s}} dx .$$

The theorem allows us to make use of the weighted-A_1 boundedness of T on L^p as follows. Let C_δ be as in part (i) of the theorem with $\delta = 1/s$, and let C' correspond to $C = C_\delta$ as in the hypothesis of the proposition. Then

$$\int_{\mathbb{R}^n} |Tf|^q dx \leq C' \int_{\mathbb{R}^n} |f|^p [((|Tf|^{(q-p)s})^*]^{\frac{1}{s}} dx \quad .$$

Apply Hölder's inequality with exponents $\frac{q}{p}$ and $\frac{q}{q-p}$ to obtain

$$\int_{\mathbb{R}^n} |Tf|^q dx \leq C(T,s) [\int_{\mathbb{R}^n} |f|^q dx]^{\frac{p}{q}} \cdot [\int_{\mathbb{R}^n} [(|Tf|^{(q-p)s})^*]^{\frac{q}{s(q-p)}} dx]^{1-\frac{p}{q}}$$

If we choose $s > 1$ so that $\frac{q}{s(q-p)} > 1$, then from the boundedness of the maximal function on $L^r(\mathbb{R}^n, dx)$ for $r > 1$ we conclude that

$$\|Tf\|_q \leq C(T,s) \|f\|_q \quad ,$$

as desired.

The interest in this proposition lies in the fact that the A_1-weighted inequalities for p imply the unweighted L^q-inequalities for $\infty > q > p$.

Let us now introduce the A_p weights for $1 < p < \infty$.

Definition. A weight ω is an A_p weight if there exists $C > 0$ (which we denote by $C_p(\omega)$) such that for all cubes Q,

$$m_Q(\omega)[m_Q(\omega^{-\frac{1}{p-1}})]^{p-1} \leq C \; .$$

Observe that this is equivalent to condition (2) in Section I above when $u = v = \omega$. Notice that $A_1 \subseteq A_p$ for $1 < p < \infty$ and, more generally, $A_{p_1} \subsetneq A_{p_2}$ when $p_2 > p_1$. In fact, this inclusion is strict: $|x|^\alpha \in A_p$ if and only if $-n < \alpha < n(p-1)$. Note also that $\omega \in A_p$ if and only if $\omega^{-\frac{1}{p-1}} \in A_{p'}$, $\frac{1}{p} + \frac{1}{p'} = 1$.

From Section I it follows that $\omega \in A_p$ if and only if the Hardy-Littlewood maximal operator is bounded from $L^p(\omega dx)$ into weak-$L^p(\omega dx)$. However, in our current situation, this may be improved.

Theorem. (Muckenhoupt [M]). The operator $f \to f^*$ is bounded on $L^p(\omega dx)$ if and only if $\omega \in A_p$.

The "only if" part of this theorem follows from our earlier results; the "sufficiency" is a consequence of the following: see [CF].

Lemma. If $p > 1$ and $\omega \in A_p$, then there exists $\epsilon(\omega) > 0$ such that $\omega \in A_{p-\epsilon}$.

Let us assume the lemma for the moment and see how this is used to provide a simple proof of the second half of the theorem. Suppose that $\omega \in A_p$. Then $\omega \in A_{p-\epsilon}$ for some $\epsilon > 0$, and hence, by our earlier results, the maximal operator takes $L^{p-\epsilon}(\omega dx)$ into weak-$L^{p-\epsilon}(\omega dx)$ boundedly. Since the maximal operator is bounded on L^∞, the theorem above now follows from the Marcinkiewicz interpolation theorem.

The lemma is itself a consequence of the following:

Theorem (The reverse Hölder inequality) Let $p \geq 1$ and suppose that $\omega \in A_p$. Then there exists $\eta > 0$ and $C(\omega,\eta) > 0$ such that

$$m_Q(\omega^{1+\eta}) \leq C(\omega,\eta)(m_Q(\omega))^{1+\eta} .$$

If we use this for the weight $\omega^{-\frac{1}{p-1}} \in A_{p'}, (\frac{1}{p} + \frac{1}{p'} = 1)$, then we obtain that

$$[m_Q(\omega^{-\frac{1}{p-1}})^{1+\eta}]^{\frac{p-1}{1+\eta}} \leq C[m_Q\omega^{-\frac{1}{p-1}}]^{p-1} .$$

Choose $\epsilon = \epsilon(\omega) > 0$ so that $\frac{p-1}{1+\eta} = p - \epsilon - 1$. Then the preceeding inequality together with the A_p condition on ω implies $\omega \in A_{p-\epsilon}$. Hence the lemma follows from the reverse Hölder inequality.

The A_p condition is itself equivalent to a reverse Hölder inequality with respect to the measure ωdx and the weight $\frac{1}{\omega}$. Indeed, if $\omega(Q) = \int_Q \omega dx$ then the A_p condition becomes

$$\left(\frac{\omega(Q)}{|Q|}\right)\left(\frac{\int_Q \omega^{-\frac{1}{p-1}}dx}{|Q|}\right)^{p-1} \leq C ,$$

or,

$$\left(\frac{\omega(Q)}{|Q|}\right)^p\left(\frac{\int_Q \omega^{-\frac{1}{p-1}}dx}{\omega(Q)}\right)^{p-1} \leq C .$$

But, this is the same thing as

$$\left[\frac{\int_Q \omega^{-\frac{p}{p-1}}\omega dx}{\int_Q \omega dx}\right]^{\frac{p-1}{p}} \leq C\,\frac{\int_Q \omega^{-1}\omega dx}{\int_Q \omega dx} ,$$

which is a reverse Hölder inequality.

Suppose now $\omega \in A_p$, so that we are left with showing that ω satisfies such a reverse Hölder condition. Assume, for the moment, that we have shown ωdx to be a doubling measure. Then, as we have remarked before, our work in this chapter still goes through if we replace Lebesgue measure by ωdx. In particular, the classes $A_p(\omega dx)$ can be defined as they were above when $\omega \equiv 1$.

Suppose that we can prove also that $\frac{1}{\omega}$ lies in $A_{p_1}(\omega dx)$ for some p_1, if $\omega \in A_p$. Then the above calculation (which showed that the A_p condition is equivalent to a reverse Hölder inequality for a slightly different measure and weight) now shows that $\frac{1}{\omega} \in A_{p_1}(\omega dx)$ is equivalent to the weight ω satisfying a reverse Hölder inequality with respect to dx. This is what we want.

Thus we must show that if $\omega \in A_p$, then ωdx is a doubling measure, and $\frac{1}{\omega} \in A_{p_1}(\omega dx)$ for some p_1.

Before continuing with our proof, we remark that this is not the quickest way of proving the reverse Hölder inequality. It does, however, emphasize the fact that the reverse Hölder condition and the A_p condition are equivalent in a very precise sense, and it provides us with an opportunity to introduce an important equivalence relation between doubling measures.

For any doubling measure $d\mu$ we define, for $1 < p < \infty$, the class $\mathcal{a}_p(d\mu)$ as the set of measures of the form $\omega d\mu$ for which there is a $C > 0$ such that, given any cube Q,

$$(\mu_Q(\omega))(\mu_Q(\omega^{-\frac{1}{p-1}}))^{p-1} \leq C .$$

We let

$$\mathcal{a}_\infty(d\mu) = \bigcup_{p>1} \mathcal{a}_p(d\mu) .$$

<u>Theorem</u>: <u>Let</u> μ <u>be a doubling measure</u>. <u>Then</u> $\nu \in \mathcal{a}_\infty(d\mu)$ <u>implies that</u> ν <u>is also doubling</u>. <u>Furthermore</u>, "$\nu \in \mathcal{a}_\infty(d\mu)$" <u>is an equivalence relation between doubling measures</u>.

In particular, if $\omega \in A_p$, then $\omega dx \in \mathcal{a}_\infty(dx)$; hence, the theorem tells us that ωdx is a doubling measure and dx $(= \frac{1}{\omega}\omega dx) \in \mathcal{a}_\infty(\omega dx)$, so that $\frac{1}{\omega} \in A_{p_1}(\omega dx)$ for some p_1. Hence the reverse Hölder inequality for A_p weights now follows from this theorem and our earlier remarks. (One should recall that we already reduced the previous theorem and its lemma to proving the reverse Hölder inequality for A_p weights).

We shall obtain the last theorem from the following:

Lemma. Let μ be a doubling measure, and suppose that $f \in L^1_{loc}(d\mu)$. Then $\omega d\mu \in \mathcal{A}_\infty(d\mu)$ if and only if there exists $0 < \alpha < 1$ and $0 < \beta < 1$ such that for any cube Q and any measurable subset E of Q ,

$$\frac{\int_E \omega d\mu}{\int_Q \omega d\mu} \leqq \beta$$

implies that

$$\frac{\int_E d\mu}{\int_Q d\mu} \leqq \alpha .$$

To see how the theorem can be obtained from the lemma, suppose that μ is a measure and $\nu = \omega d\mu \in \mathcal{A}_\infty(d\mu)$. Then, by the lemma, this is equivalent to

$$\frac{\int_E \omega d\mu}{\int_Q \omega d\mu} \leqq \beta \quad \text{implies} \quad \frac{\int_E d\mu}{\int_Q d\mu} \leqq \alpha$$

for all Q and all measurable $E \subseteq Q$, or equivalently, (replacing E by its relative complement $Q - E$)

$$\frac{\int_E d\mu}{\int_Q d\mu} \geqq 1 - \alpha \quad \text{implies} \quad \frac{\int_E \omega d\mu}{\int_Q \omega d\mu} \geqq 1 - \beta .$$

From this it is easy to see that $\omega d\mu$ will be doubling if μ is, and in that case, $\mathcal{A}_\infty(\omega d\mu)$ makes sense. The above calculation and the lemma clearly show that "$\omega d\mu \in \mathcal{A}_\infty(d\mu)$" is equivalent to "$d\mu \in \mathcal{A}_\infty(\omega d\mu)$". Thus the relation we introduced is symmetric. That it is transitive will follow immediately from a slightly different version of this lemma which is a consequence of its proof, and is stated as a "proposition" below. Hence, "$\nu \in \mathcal{A}_\infty(d\mu)$" is an equivalence relation.

Proof of the lemma: We consider the necessity portion of the lemma first. As usual all of the arguments given will extend, by a mere change of notation, from the case $d\mu = dx$ to the general case of μ a doubling measure, and so we assume that μ is Lebesgue measure.

Suppose that $\omega \in A_p$ for some $p \geqq 1$ (i.e., $\omega dx \in \mathcal{A}_\infty(dx)$) . Let Q be a cube and E a measurable subset of Q . Then

$$Q \subseteq \{\chi_E^*(x) \geqq \frac{|E|}{|Q|}\} ,$$

so that

$$\int_Q \omega dx \leqq \int_{\{\chi_E^*(x) \geqq \frac{|E|}{|Q|}\}} \omega dx \leqq \frac{C}{\left(\frac{|E|}{|Q|}\right)^p} \int_{R^n} |\chi_E|^p \omega dx ,$$

since $\omega \in A_p$ implies that the maximal function takes $L^p(\omega dx)$ to weak-$L^p(\omega dx)$ boundedly. (See Section I.)

This implies that $\left(\frac{|E|}{|Q|}\right)^p \leqq C\left(\frac{\omega(E)}{\omega(Q)}\right)$, which gives us the desired conclusion, (and this is, in fact, equivalent to the condition in the lemma, as follows from the sufficiency part of the lemma).

To prove sufficiency, let $0 < \alpha < 1$ and $0<\beta<1$ be as above, so that ωdx is a doubling measure. Suppose that we are given a cube Q_0, and let $\Delta(Q_0)$ be the system of dyadic cubes in R^n relative to Q_0. (See Section II of the preceeding chapter.)

Let $A > 1$ be a constant to be chosen later. For each $k \geqq 1$, we apply the Calderón-Zygmund decomposition to the function $\frac{1}{\omega} \chi_{Q_0}$ relative to the measure ωdx, for the value $A^k|Q_0| / \omega(Q_0)$ and the dyadic system $\Delta(Q_0)$. Thus, we obtain a family $\{Q_k^i\}_{i \in I}$ of cubes in $\Delta(Q_0)$ (that are maximal with respect to inclusion in a certain open set-see Section II of the last chapter) which have the following properties:

(i) $$A^k \frac{|Q_0|}{\omega(Q_0)} < \frac{|Q_0 \cap Q_i^k|}{\omega(Q_i^k)} \leqq C_\omega A^k \frac{|Q_0|}{\omega(Q_0)} \quad ;$$

(ii) $$\frac{1}{\omega(x)} \leqq A^k \frac{|Q_0|}{\omega(Q_0)} \quad \text{a.e. on} \quad (\bigcup_{i \in I} Q_k^i)^c \ .$$

Note that all of the cubes Q_k^i must be contained in Q_0, since $A > 1$.

Let $D_k = \bigcup_{i \in I} Q_k^i$.

Claim. Choose $A = \frac{C_\omega}{\beta}$, with C_ω the constant in (i) above. Then

$$\frac{\omega(Q_i^k \cap D_{k+1})}{\omega(Q_i^k)} \leqq \beta$$

for each i and k.

To prove the claim, we begin by observing that each Q_j^{k+1} is contained in some Q_i^k, as one can easily verify by looking at the construction used in the Calderón-Zygmund decomposition. From (i) above we obtain that

$$\omega(Q_j^{k+1}) \leqq \frac{1}{A^{k+1}} \frac{\omega(Q_0)}{|Q_0|} |Q_j^{k+1}|$$

so that

$$\omega(Q_i^k \cap D_{k+1}) \leqq \frac{1}{A^{k+1}} \frac{\omega(Q_0)}{|Q_0|} |Q_i^k| \ .$$

Hence, again by (i),

$$\leqq \frac{1}{A^{k+1}} C_\omega A^k \omega(Q_i^k)$$

$$\leqq \beta\omega(Q_i^k) \ ,$$

as desired.

If we now apply the hypothesis to the conclusion of the claim, we obtain

$$|Q_i^k \cap D_{k+1}| \leqq \alpha |Q_i^k| \ .$$

Because $D_{k+1} \subseteq D_k$ (again, by the construction used in the Calderón-Zygmund decomposition), summing over i gives

$$|D_{k+1}| \leqq \alpha |D_k| \ ,$$

or

$$|D_k| \leqq \alpha^k |D_0| \leqq \alpha^k |Q_0| \ .$$

In order to prove that $\omega \in A_p$ for some p (that is, $\omega dx \in G_p(dx)$ for some p), it suffices to show that the weight $\frac{1}{\omega}$ satisfies a reverse Hölder inequality with respect to the measure ωdx. (Before the statement of the last theorem we observed that these two conditions are equivalent.)

Thus, let $\delta > 0$ (to be specified later). Then,

$$\int_Q \frac{1}{\omega^{1+\delta}} \omega dx$$

$$\leqq \int_{Q \setminus D_0} \frac{\omega}{\omega^{1+\delta}} dx + \left[\sum_{k=0} \int_{D_k \setminus D_{k+1}} \frac{\omega}{\omega} dx \left(A^{k+1} \frac{|Q_0|}{\omega(Q_0)} \right)^\delta \right]$$

(using (ii) above for both terms)

$$\leq |Q_0| \left[\frac{|Q_0|}{\omega(Q_0)}\right]^{\delta} + \sum_{k=0}^{\infty} |D_k| \left(A^{k+1} \frac{|Q_0|}{\omega(Q_0)}\right)^{\delta}$$

$$\leq |Q_0| \left[\frac{|Q_0|}{\omega(Q_0)}\right]^{\delta} \left(1 + A^{\delta} \sum_{k=0}^{\infty} (\alpha A^{\delta})^k\right) .$$

If we choose $\delta > 0$ sufficiently small so that $\alpha A^{\delta} < 1$, then this latter series converges, and we obtain

$$\left[\frac{\omega(Q_0)}{|Q_0|}\right] \left[m_{Q_0} \omega^{-\delta}\right]^{\frac{1}{\delta}} \leq C .$$

Because all of our constants are independent of Q_0, we conclude that $\omega \in A_{1+\frac{1}{\delta}}$, which finishes the proof of the lemma.

We also have the following.

Proposition. Let μ be a doubling measure, and suppose that $\omega \in L^1_{loc}(d\mu)$. Then $\omega d\mu \in \mathcal{C}_\infty(d\mu)$ if and only if there exist positive constants $C_1, C_2, \delta_1, \delta_2$ such that for any cube Q and any measurable subset E of Q,

$$C_1 \left(\frac{\mu(E)}{\mu(Q)}\right)^{\delta_1} \leq \frac{\int_E \omega d\mu}{\int_Q \omega d\mu} \leq C_2 \left(\frac{\mu(E)}{\mu(Q)}\right)^{\delta_2} .$$

The sufficiency part of this proposition follows from the corresponding portion of the lemma, whereas necessity follows from the first part of the proof of the lemma. Note that the transitivity of the relation "$\nu \in \mathcal{C}_\infty(d\mu)$" follows immediately from this proposition, as announced earlier.

We can now use the reverse Hölder inequality for A_1 weights to prove the second part of the Coifman-Rochberg theorem.

Let Q be a cube and let ω be an A_1 weight, so that ω satisfies the reverse Hölder condition. That is, there exist $\varepsilon > 0$ and $C > 0$, independent of Q, such that

$$(m_Q \omega^{1+\varepsilon})^{\frac{1}{1+\varepsilon}} \leq C m_Q \omega ,$$

so that

$$(m_Q \omega^{1+\varepsilon})^{\frac{1}{1+\varepsilon}} \leq C \operatorname*{ess\,inf}_{x \in Q} \omega(x) ;$$

hence,

$$[(\omega^{1+\epsilon})^*]^{\frac{1}{1+\epsilon}} \leqq C\omega \quad \text{a.e.}$$

Because $\omega^* \geqq \omega$ a.e. we choose $\delta = \frac{1}{1+\epsilon}$ and $f = \omega^{1+\epsilon}$, so that the theorem follows, for an appropriate choice of h.

We conclude this chapter by stating an extrapolation result of José Rubio de Francia, (which can be stated more generally). (See [G] and [R].)

Theorem: Let $p_0 \geqq 1$, and suppose that T is a linear operator which is bounded on $L^{p_0}(\omega dx)$ for all $\omega \in A_{p_0}$, with the operator norm controlled by the A_{p_0} constant of ω. Then the same is true for any p, $1 < p < \infty$.

For an example of an application, (see the end of Chapter 6).

CHAPTER 3

$BMO(R^n, dx)$

I. Definition and Basic Properties.

Let $f \in L^1_{loc}(R^n)$ be such that $\exp f \in A_2$. Then

$$(m_Q \exp f)(m_Q \exp(-f)) \leqq C$$

so that

$$(m_Q \exp(f - m_Q f))(m_Q \exp(m_Q f - f)) \leqq C \ .$$

By Jensen's inequality, each factor is at least one, and hence each is no greater than C. Thus,

$$m_Q \exp|f - m_Q f| \leqq 2C \ ;$$

hence,

$$m_Q|f - m_Q f| \leqq 2C \ .$$

Definition: For $f \in L^1_{loc}(R^n)$ we let $\|f\|_* = \sup_Q m_Q|f - m_Q f|$, and we define the space $BMO(R^n)$ (of functions of "bounded mean oscillation") to consist of those functions f such that $\|f\|_* < \infty$.

$(BMO(R^n), \| \ \|_*)$ is a semi-normed vector space, with the seminorm vanishing on the constant functions. If we let $\mathbb{C}$ denote the vector space of constant functions, then the quotient of $BMO(R^n)$ by $\mathbb{C}$ is a Banach space, which we also denote by $BMO(R^n)$. It will always be understood in a given context whether an element of $BMO(R^n)$ is to be a function or an equivalence class of functions modulo constants.

This space BMO was originally introduced by John and Nirenberg in [JN].

Coifman and Rochberg's theorem (of the second section in the preceeding chapter) and the preceeding calculation provide us with a large class of examples of BMO functions: there exists $C(n) > 0$ such that if $f \in L^1_{loc}(R^n)$ (and $f^* \not\equiv \infty$) then $\|\log f^*\|_* \leqq C(n)$. In particular, one can use this to show that $\log|x| \in BMO$.

To estimate $\|f\|_*$, it is sometimes useful to notice that for any $\alpha \in \mathbb{C}$

$$\frac{1}{|Q|}\int_Q |f - m_Q f|\,dx \leqq \frac{2}{|Q|}\int_Q |f - \alpha|\,dx \ .$$

From this we see that

$$\|f\|_*' = \sup_Q \inf_{\alpha \in C} m_Q|f - \alpha|$$

defines an equivalent norm on $BMO(R^n)$.

When f is real valued and Q is any cube, the constants α_Q for which $\inf_{\alpha \in R} m_Q|f - \alpha|$ is attained are the ones which satisfy

$$|\{x \in Q : f > \alpha_Q\}| \leqq \frac{1}{2}|Q|$$

and

$$|\{x \in Q : f < \alpha_Q\}| \leqq \frac{1}{2}|Q| \quad .$$

Even if it is not unique, we call α_Q "the good constant" .

As we shall see later, BMO is a good substitute space for L^∞ . Among the common features of L^∞ and BMO are the existence of good control on the size of the functions and the nonexistence of any subspace of "nice" functions which is dense in the norm topology. Fortunately, however, approximation of BMO functions by bounded functions in the topology of L^1_{loc} is very easy to do, and in pratice, usually sufficient.

<u>Lemma</u>: <u>Let</u> $f \in BMO(R^n)$, <u>and for</u> $q \in N$ <u>define</u>

$$f_q : R^n \to C \quad \underline{\text{by}}$$

$$f_q(x) = f(x) \quad \underline{\text{if}} \quad |f(x)| \leqq q \ ,$$

<u>and</u>

$$f_q(x) = f(x) \frac{q}{|f(x)|} \quad \text{if} \quad |f(x)| > q \quad .$$

<u>Then</u>

$$\|f_q\|_*' \leqq C\|f\|_*' \ .$$

This lemma is immediate, as are the following properties of the sequence $\{f_q\}$:

(i) $\|f_q\|_\infty \leqq q$;

(ii) $f_q \to f$ in $L^1_{loc}(R^n, dx)$;

(iii) $|f_q - f|$ tends monotonically to zero.

We should also point out that the lemma holds if f takes its values in some Banach space B , where $BMO_B(R^n)$ is defined appropriately. Since the only facts used are:

(i) for each cube Q and $f \in BMO(R^n)$, and for $q > 0$,

$$\inf_{\alpha\in B} m_Q|f_q-\alpha| \leq C \inf_{\alpha\in B, \|\alpha\|_B\leq q} m_Q|f_q-\alpha| \ ;$$

(ii) if $\alpha,\beta\in B$ and $\|\alpha\|_B \leq q < \|\beta\|_B$, then

$$\|\alpha - q\frac{\beta}{\|\beta\|_B}\|_B \leq C\|\alpha-\beta\|_B \ .$$

In particular, (i) is a consequence of (ii).

The following result gives one of the main properties of BMO functions, and, in particular, it gives a very strong local control on their size. In its proof we see an example of how the preceeding lemma can be used.

Theorem: (Inequality of John and Nirenberg) There exist two positive constants $\lambda>0$ and $C>0$ such that for any $f\in BMO(\mathbb{R}^n)$,

$$\sup_Q \frac{1}{|Q|}\int_Q \exp\left(\frac{\lambda}{\|f\|_*}|f-m_Qf|\right)dx \leq C \ .$$

Proof: We assume that f is bounded, so that the above supremum makes sense for all λ, and we shall prove the theorem by finding a bound independent of $\|f\|_\infty$.

Let Q_0 be a fixed cube and Q some cube (in the dyadic "mesh") $\Delta(Q_0)$. Recall that $\tilde{Q}$ is then the unique element of $\Delta(Q_0)$ which contains Q and lies in "the previous generation".

Lemma: $|m_Qf - m_{\tilde{Q}}f| \leq 2^n\|f\|_*$.

Indeed, $|m_Qf - m_{\tilde{Q}}f| \leq m_Q|f-m_{\tilde{Q}}f| \leq 2^n m_{\tilde{Q}}|f-m_{\tilde{Q}}f| \leq 2^n\|f\|_*$.

Consider now the Calderón-Zygmund decomposition of the function $(f-m_{Q_0}f)\chi_{Q_0}$ for $\lambda = 2\|f\|$. This yields a collection of dyadic cubes Q_i, maximal with respect to inclusion, satisfying

$$m_{Q_i}|(f-m_{Q_0}f)\chi_{Q_0}| > 2\|f\|_*$$

and

$$|(f-m_{Q_0}f)\chi_{Q_0}| \leq 2\|f\|_* \quad \text{on} \quad (\cup Q_i)^c \ .$$

Clearly, $Q_i \subseteq Q_0$ for each i, and

$$|\cup Q_i| \leq \frac{\|(f-m_{Q_0}f)\chi_{Q_0}\|_1}{2\|f\|_*} \leq \frac{|Q_0|}{2} \ .$$

Since the Q_i's are maximal, $m_{\tilde{Q}_i}|f - m_{Q_0}f| \leq 2\|f\|_*$, and the previous lemma gives

$$|m_{Q_i}f - m_{Q_0}f| \leq |m_{Q_i}f - m_{\tilde{Q}_i}f| + |m_{\tilde{Q}_i}f - m_{Q_0}f| \leq (2^n+2)\|f\|_* .$$

Let $X(\lambda) = \sup_Q \frac{1}{|Q|}\int_Q \exp(\frac{\lambda}{\|f\|_*}|f - m_Q f|)dx$, which is finite, since we are assuming that f is bounded. From the properties of the Q_i we obtain that

$$\frac{1}{|Q_0|}\int_{Q_0}\exp(\frac{\lambda}{\|f\|_*}|f - m_{Q_0}|)dx$$

$$\leq \frac{1}{|Q_0|}\int_{Q_0\setminus \bigcup_i Q_i} e^{2\lambda}dx + \frac{1}{|Q_0|}\sum_i \frac{|Q_i|}{|Q_i|}\left[\int_{Q_i}\exp(\frac{\lambda}{\|f\|_*}|f - m_{Q_i}f|)dx\, e^{(2^n+2)\lambda}\right]$$

$$\leq e^{2\lambda} + \frac{1}{2}[\exp((2^n+2)\lambda)]X(\lambda) .$$

From taking the supremum over all cubes Q_0 it follows that

$$X(\lambda)[1 - \frac{1}{2}\exp((2^n+2)\lambda)] \leq e^{2\lambda} ,$$

which implies that $X(\lambda) \leq C$, if λ is small enough, which proves the theorem.

A consequence of the theorem, which in fact is equivalent to it, is the following: There exist positive constants λ and C such that for every cube Q and every $t \in R_+$,

$$|\{x \in Q : |f(x) - m_Q f| > t\|f\|_*\}| \leq C e^{-\lambda t}|Q| .$$

For $1 \leq p < \infty$ let

$$\|f\|_{p,*} = \sup_Q [m_Q(|f - m_Q f|^p)]^{1/p} .$$

Then, perhaps one of the most important consequences of the John-Nirenberg inequality if that these all define equivalent norms on BMO . Indeed, since $x^p \leq C(p,\lambda)e^{\lambda x}$ for $x > 0$, it clearly follows from the theorem that $\|f\|_{p,*} \leq C(p)\|f\|_*$. The converse inequality follows from a simple application of Hölder's inequality.

Another application of the John-Nirenberg inequality is the following.

<u>Proposition</u>: <u>For</u> $1 < p < \infty$ <u>define</u> $\mathcal{U}_p$ <u>by</u>

$$\mathcal{U}_p = \{b \in \mathrm{BMO} : e^b \in A_p\} .$$

Then $\mathcal{U}_p$ is an open subset of BMO.

Proof: Let $\omega \in A_p$, so that $\omega^{-\frac{1}{p-1}} \in A_{p'}$. Then the reverse Hölder inequality for weights applied to both ω and $\omega^{-\frac{1}{p-1}}$ shows that $\omega^{1+\epsilon} \in A_p$, if $\epsilon > 0$ is sufficiently small.

We would like to show that $\omega e^b \in A_p$ when $\|b\|_*$ is small enough. By Hölder's inequality,

$$\left[\frac{1}{|Q|}\int_Q \omega e^b dx\right]\left[\frac{1}{|Q|}\int_Q (\omega e^b)^{-\frac{1}{p-1}} dx\right]^{p-1}$$

$$\leq \left[\frac{1}{|Q|}\int_Q \omega^{1+\epsilon} dx\right]^{\frac{1}{1+\epsilon}}\left[\frac{1}{|Q|}\int_Q (\omega^{1+\epsilon})^{-\frac{1}{p-1}}\right]^{\frac{1}{(1+\epsilon)(p-1)}}\left[\frac{1}{|Q|}\int_Q e^{\frac{1+\epsilon}{\epsilon}b} dx\right]^{\frac{\epsilon}{1+\epsilon}}\left[\frac{1}{|Q|}\int_Q e^{-\frac{1}{p-1}\frac{1+\epsilon}{\epsilon}b} dx\right]^{(1+\epsilon)(p-1)}$$

The first two factors are controlled by the fact that $\omega^{1+\epsilon} \in A_p$. If we multiply the second pair by $e^{-m_Q b} e^{m_Q b}$, and apply the John-Nirenberg inequality for $\|b\|_*$ small enough, then the last two factors are also appropriately controlled.

The following result is a natural complement to the previous theorem, in that it gives information on the size of a BMO function at infinity.

Lemma: Let $\epsilon > 0$. Then there exists $C(\epsilon) > 0$ such that for any cube Q_0, with side length δ and center x_0, and any $f \in \mathrm{BMO}(\mathbb{R}^n)$,

$$\int_{\mathbb{R}^n} \frac{|f(x) - m_{Q_0} f|}{\delta^{n+\epsilon} + |x - x_0|^{n+\epsilon}} dx \leq \frac{C(\epsilon)}{\delta^\epsilon} \|f\|_* .$$

Proof: We employ the common trick of decomposing $\mathbb{R}^n$ into a geometrically increasing sequence of concentric cubes. Let $Q_k = 2^k Q_0$, so that $Q_{k+1} = \overline{Q_k}$. Then, as in the lemma following the John-Nirenberg theorem,

$$|m_{Q_{k+1}} f - m_{Q_k} f| \leq 2^n \|f\|_* .$$

Hence,

$$|m_{Q_k} f - m_{Q_0} f| \leq k\, 2^n \|f\|_* .$$

Thus, setting $Q_{-1} = \emptyset$ for convenience,

$$\int_{R^n} \frac{|f(x) - m_{Q_0}|}{\delta^{n+\epsilon} + |x - x_0|^{n+\epsilon}} \, dx$$

$$\leq \sum_{n=0}^{\infty} \int_{Q_k - Q_{k-1}} \frac{|f(x) - m_{Q_0} f|}{\delta^{n+\epsilon} + |x - x_0|^{n+\epsilon}} \, dx$$

$$\leq C \sum_{n=0}^{\infty} \int_{Q_k} \frac{|f(x) - m_{Q_0} f|}{(2^k \delta)^{n+\epsilon}} \, dx$$

$$\leq C \sum_{n=0}^{\infty} \frac{1}{(2^k \delta)^{\epsilon}} m_{Q_k} |f - m_{Q_0} f|$$

$$\leq C \sum_{k=0}^{\infty} \frac{1}{(2^k \delta)^{\epsilon}} (k+1) 2^n \|f\|_* \leq \frac{C(\epsilon)}{\delta^{\epsilon}} \|f\|_* \, .$$

The same sort of calculation also proves the following lemma, which will be used frequently in the following chapters.

Lemma: Let $\epsilon > 0$. Then there exists $C(\epsilon) > 0$ such that for each cube $Q_0 = Q(x_0, \delta)$ and each $f \in L^1_{loc}(R^n)$,

$$\int_{R^n} \frac{|f(x)| \delta^{\epsilon}}{\delta^{n+\epsilon} + |x - x_0|^{n+\epsilon}} \, dx \leq C(\epsilon) \inf_{x \in Q_0} f^*(x) \, .$$

II. $BMO(R^n)$ Viewed as a Dual Space.

In [FS], Feffermann and Stein have characterized $BMO(R^n, dx)$ as the dual of the space $H^1(R^n)$. $BMO(R^n, dx)$ can also be viewed very naturally as the dual of an "atomic" space $H^{1,\infty}(R^n)$ defined below. To see this we shall make a slight change on the definition of BMO-norm of a function f. For the time being, all functions are assumed to be real valued.

Let Q be any cube in R^n, and denote by $\mathfrak{a}_Q^{\infty}$ the set of functions a such that

(i) $a \in L^{\infty}(R^n)$, $\|a\|_{\infty} \leq \frac{1}{|Q|}$;

(ii) $\operatorname{supp} a \subseteq Q$;

(iii) $m_Q(a) = 0$.

Let $f \in L^1_{loc}(\mathbb{R}^n)$ and let α_Q be the "good constant" for f on Q (which is defined even if f is not in BMO). Then

$$\frac{1}{|Q|}\int_Q |f-\alpha_Q|\,dx = \sup_{a \in \mathcal{A}^\infty} \left|\int_{\mathbb{R}^n} af\,dx\right| .$$

That the left side dominates the right is trivial. To see the opposite inequality, we define $a \in \mathcal{A}^\infty_Q$ be setting $a = \frac{1}{|Q|}$ on $\{x \in Q : f(x) > \alpha_Q\}$, $a = \frac{-1}{|Q|}$ on $\{x \in Q : f(x) < \alpha_Q\}$, and letting a , on $\{x \in Q : f(x) = \alpha_Q\}$, be such that $|a| = \frac{1}{|Q|}$ and (iii) (above) is satisfied. This may be done, since

$$|\{x \in Q : f(x) < \alpha_Q\}| \leqq \frac{1}{2}|Q|$$

and

$$|\{x \in Q : f(x) > \alpha_Q\}| \leqq \frac{1}{2}|Q| .$$

Clearly, then,

$$\frac{1}{|Q|}\int_Q |f-\alpha_Q|\,dx = \left|\int_{\mathbb{R}^n} f\,a\,dx\right| .$$

Let $\mathcal{A}^\infty = \bigcup_Q \mathcal{A}^\infty_Q$. The functions in $\mathcal{A}^\infty$ are called $(1,\infty)$-atoms, and they were introduced by Coifman and Weiss in a much more general context. See [CW1].

All of the functions a in $\mathcal{A}^\infty$ satisfy $\|a\|_1 \leqq 1$. Thus, we can define a Banach space $H^{1,\infty}$ ("atomic H^1 ") which is continuously included in $L^1(\mathbb{R}^n, dx)$ in the following way:

$f \in H^{1,\infty}$ if and only if there exists $\lambda_i \in \mathbb{R}$ and $a_i \in \mathcal{A}^\infty$ such that $\Sigma|\lambda_i| < \infty$ and $f = \sum_i \lambda_i a_i$. For $f \in H^{1,\infty}$, we define its norm $\|f\|_{H^{1,\infty}}$ to be $\inf(\Sigma|\lambda_i|)$, where the infimum is taken over all sequences $(\lambda_i)_{i \in I}$ occuring in such an "atomic decomposition" of f .

Theorem: $(H^{1,\infty})^* = BMO$.

For $b \in BMO$, define $L(b)$ on $\mathcal{A}^\infty$ by

$$L(b)(a) = \int_{\mathbb{R}^n} b\,a\,dx .$$

Then $L(b)$ is bounded on $\mathcal{A}^\infty$, and hence it extends in a natural way to a bounded linear functional on $H^{1,\infty}$, that is , $L(b) \in (H^{1,\infty})^*$. The conclusion of the theorem is that every element of $(H^{1,\infty})^*$ arises in this manner, and the norm of $L(b)$ as a functional is equivalent to the BMO norm of b.

It is not hard to show that L is an isomorphism of BMO onto $L(\mathrm{BMO})$, that is, that there exist $C_1 > 0$ and $C_2 > 0$ such that

$$C_1 \leq \frac{\|b\|_*}{\|L(b)\|_{(H^{1,\infty})^*}} \leq C_2$$

for all $b \in \mathrm{BMO}$. In fact, it is true that

$$\|b\|'_* = \|L(b)\|_{(H^{1,\infty})^*} \leq C_2$$

(see page 34 for the definition of $\| \;\|'_*$.)

We first note that $\|b\|'_* \geq \|L(b)\|_{(H^{1,\infty})^*}$, since, by the calculation above, $\|b\|'_* \geq |\int_{R^n} badx|$ if $a \in \mathcal{a}^\infty$.

Let Q be any cube, and let $a_0 \in \mathcal{a}^\infty_Q$ be the atom constructed near the beginning of this section which satisfies

$$\inf_{\alpha \in C} \int_Q |b - \alpha| \frac{dx}{|Q|} = |\int_{R^n} ba_0 dx| .$$

This a_0 also satisfies $|a_0| = \frac{1}{|Q|}$ on Q, so that $1 \geq \|a_0\|_{H^{1,\infty}} \geq \|a_0\|_1 = 1$ (since for any $f \in H^{1,\infty}$, $\|f\|_{H^{1,\infty}} \geq \|f\|_1$). Thus,

$$\|b\|'_* \leq \sup_{\|a\|_{H^{1,\infty}} = 1} |\int_{R^n} badx| = \|L(b)\|_{(H^{1,\infty})^*} ,$$

as desired.

To prove the theorem, what remains to be shown is that $L(\mathrm{BMO}) = (H^{1,\infty})^*$.

We shall consider a family of "atomic" spaces $H^{1,p}$ for $1 < p < \infty$ whose duals can be canonically realized (via L above) as BMO, and such that H^{1,p_2} is continuously included in H^{1,p_1} if $p_1 \leq p_2 \leq \infty$. Once we have done this, the theorem will follow from an application of Banach's closed range theorem. (See [Y].)

Indeed, consider the following diagram:

$$\begin{array}{ccc} H^{1,\infty} & \xrightarrow{\ i\ } & H^{1,p} \\ (H^{1,\infty})^* & \xleftarrow{\ i^*\ } & \mathrm{BMO} = (H^{1,p})^* \end{array} .$$

Since i is an inclusion, i^* is the canonical injection of BMO in $(H^{1,\infty})^*$, which is L. Because $L(\mathrm{BMO})$ is closed in $(H^{1,\infty})^*$, the aforementioned theorem of Banach implies that $H^{1,\infty}$ is closed in $H^{1,p}$. The Hahn-Banach theorem and the fact that $(H^{1,p})^* = \mathrm{BMO}$ imply then that $H^{1,\infty} = H^{1,p}$ for $1 < p < \infty$. (This could also be obtained by using the definition of $H^{1,p}$ below to show directly that $H^{1,\infty}$ is dense in $H^{1,p}$.) Hence $(H^{1,\infty})^* = \mathrm{BMO}$.

We define $H^{1,p}$, $1 < p < \infty$, from a family of "atoms", as we did for $H^{1,\infty}$. For a cube Q let $\mathcal{a}_Q^p$ be the set of functions a such that:

(i) $a \in L^p(\mathbb{R}^n)$, $\|a\|_p \leqq |Q|^{\frac{1}{p}-1}$;

(ii) $\operatorname{supp} a \subseteqq Q$;

(iii) $m_Q(a) = 0$.

Set $\mathcal{a}^p = \bigcup_Q \mathcal{a}_Q^p$. Then it is easily verified that $\mathcal{a}^{p_2} \subseteqq \mathcal{a}^{p_1}$ if $p_1 < p_2$, and $\|a\|_1 \leqq 1$ if $a \in \mathcal{a}^p$.

We define the Banach space $H^{1,p}$ and the corresponding norm from $\mathcal{a}^p$ just as we did for $H^{1,\infty}$ and $\mathcal{a}^\infty$. H^{1,p_2} is clearly continuously embedded into H^{1,p_1} if $1 < p_1 < p_2 \leqq \infty$.

It remains to be shown that $(H^{1,p})^* = \mathrm{BMO}$ for $1 < p < \infty$. We break up the proof of this into three parts:

(1) $\mathrm{BMO} \subseteqq (H^{1,p})^*$;

(2) $(H^{1,p})^* \cap L^{p'}_{loc} \subseteqq \mathrm{BMO}$;

(3) $(H^{1,p})^* \subseteqq L^{p'}_{loc}$

To prove (1), it is enough to show that for $a \in \mathcal{a}^p$ and $b \in \mathrm{BMO}$,

$$\left|\int_{\mathbb{R}^n} ab\,dx\right| \leqq C_p \|b\|_* .$$

This follows from the equivalence on BMO of the norms $\|\ \|_*$ and $\|\ \|_{p',*}$. Indeed, if $a \in \mathcal{a}_Q^p$,

$$|\int_Q a\, b dx| = |\int_Q a(b - m_Q b)\, dx|$$

$$\leqq [\int_Q |a|^p dx]^{\frac{1}{p}} [\int_Q |b - m_Q b|^{p'} dx]^{\frac{1}{p'}}$$

$$\leqq \frac{1}{|Q|^{1-\frac{1}{p}}} [\int_Q |b - m_Q b|^{p'} dx]^{\frac{1}{p'}}$$

$$\leqq \|b\|_{p',*} \leqq C_p \|b\|_* \ .$$

To prove (2), it is enough to show that if $g \in L^{p'}_{loc}$ is given, then, for each cube Q , we can find an $a = a_Q \in H^{1,p}$ such that

$$\left[\frac{1}{|Q|}\int_Q |g - \alpha_Q|\, dx\right] \|a\|_{H^{1,p}} \leqq C_p |\int_Q g\, a dx|$$

and $\operatorname{supp} a \subseteqq Q$.

We assume, without loss in generality that,

$$\int_{Q\cap\{g>\alpha_Q\}} |g - \alpha_Q|^{p'} dx \geqq \int_{Q\cap\{g<\alpha_Q\}} |g - \alpha_Q|^{p'} dx \ ,$$

We define a by $a = 0$ outside Q ,

$$a = |g - \alpha_Q|^{p'-1} \quad \text{on} \quad Q \cap \{g > \alpha_Q\}$$

$$a = C(g,Q) \quad \text{on} \quad Q \cap \{g \leqq \alpha_Q\} \ ,$$

where $C(g,Q)$ is a constant chosen so that $m_Q a = 0$.

By the definition of α_Q ,

$$|Q \cap \{g \leqq \alpha_Q\}| \geqq \frac{1}{2}|Q| \geqq |Q \cap \{g > \alpha_Q\}| \ ;$$

hence,

$$\|a\|_{H^{1,p}} \leqq |Q|^{1-\frac{1}{p}} \|a\|_p$$

$$\leqq |Q| \left[\frac{1}{|Q|}\int_{Q\cap\{g>\alpha_Q\}} |g - \alpha_Q|^{p'} dx + \frac{1}{|Q|}\int_{Q\cap\{g\leqq\alpha_Q\}} C(g,Q)^p dx\right]^{\frac{1}{p}} .$$

But,

$$\frac{1}{|Q|}\int_{Q\cap\{g\leqq\alpha_Q\}} C(g,Q)^p dx \leqq \frac{1}{|Q\cap\{g\leqq\alpha_Q\}|}\int_{Q\cap\{g\leqq\alpha_Q\}} C(g,Q)^p dx$$

$$\leqq\left[\frac{1}{|Q\cap\{g\leqq\alpha_Q\}|}\int_{Q\cap\{g\leqq\alpha_Q\}} C(g,Q)\,dx\right]^p$$

$$\leqq\left[\frac{1}{|Q\cap\{g\leqq\alpha_Q\}|}\int_{Q\cap\{g>\alpha_Q\}} |g-\alpha_Q|^{p'-1}dx\right]^p$$

$$\leqq\left[\frac{1}{|Q\cap\{g\leqq\alpha_Q\}|}\int_{Q\cap\{g>\alpha_Q\}} |g-\alpha_Q|^{p'}\,dx\right] .$$

Thus,

$$\|a\|_{H^{1,p}} \leqq |Q|\left[\frac{3}{|Q|}\int_{Q\cap\{g>\alpha_Q\}} |g-\alpha_Q|^{p'}dx\right]^{\frac{1}{p}} .$$

To estimate $\int_Q ga\,dx$, we observe that $(g-\alpha_Q)a \geqq 0$ on Q , and hence

$$\int_Q g\,adx = \int_Q (g-\alpha_Q)adx$$

$$\geqq \int_{Q\cap\{g>\alpha_Q\}} (g-\alpha_Q)^p dx$$

$$\geqq \frac{1}{2}\int_Q |g-\alpha_Q|^{p'}\,dx .$$

From these two estimates we obtain

$$\left[\frac{1}{|Q|}\int_Q |g-\alpha_Q|^{p'}dx\right]^{\frac{1}{p'}}\|a\|_{H^{1,p}} \leqq 3^{\frac{1}{p}}\int_Q |g-\alpha_Q|^{p'}\,dx$$

$$\leqq 2\cdot 3^{\frac{1}{p}}\int_Q g\,adx .$$

In proving (3), our main tool shall be the Riesz representation theorem.

Let $\{Q_k\}$ be an increasing sequence of cubes which exhausts R^n . For each k we let T_k denote the operator which takes a function on R^n and restricts it to

Q_k , we let $L_0^p(Q_k)$ denote the subspace of $L^p(Q_k)$ consisting of functions having mean value zero, and we let $C(Q_k)$ denote the space of the functions that are constant on Q_k .

Suppose that $L \in (H^{1,p})^*$. Then

$$L \circ T_k \in [L_0^p(Q_k)]^* = L^{p'}(Q_k)/C(Q_k) .$$

Indeed, if $f \in L_0^p(Q_k)$, then

$$|L(f)| \leq \|L\|_{(H^{1,p})^*} \gamma \|f\|_{H^{1,p}} \leq \|L\|_{(H^{1,p})^*} |Q_k|^{1-\frac{1}{p}} \|f\|_p .$$

Hence, there exists a "function" g_k in $L^{p'}(Q_k)/C(Q_k)$ such that for every $f \in L_0^p(Q_k)$,

$$L(f) = \int_{Q_k} g_k f dx .$$

Now, since $L_0^p(Q_k)$ "$\subseteq$" $L_0^p(Q_{k+1})$,

$$L(f) = \int_{Q_k} g_{k+1} f dx .$$

for all $f \in L_0^p(Q_k)$. This implies that $T_k(g_{k+1}) = g_k$, and the existence of a "function" $g \in L^{p'}_{loc}(\mathbb{R}^n)/C(\mathbb{R}^n)$ such that $T_k(g) = g_k$ for all k is now automatic. Moreover, this "function" g also satisfies

$$L(f) = \int_{\mathbb{R}^n} g f dx$$

for all $f \in \mathcal{G}^p$.

This completes the proof of $(H^{1,\infty})^* = BMO$. In the next section, we shall see that the identification of $H^{1,\infty}$ with $H^{1,p}$, $1 < p < \infty$, is important in its own right. It should be noted that it is not necessary to use duality in proving this identification. (See [CW1].) In particular, it is also true for functions taking their values in some Banach space.

III. Interpolation Between $H^{1,\infty}$ and BMO.

The #-function, introduced by Fefferman and Stein, [FS], provides a tool for interpolation between L^p and BMO . It is also applicable for interpolating between

$H^{1,\infty}$ and BMO via the identification of $H^{1,\infty}$ and $H^{1,p}$.

Definition: For $f \in L^1_{loc}(R^n, dx)$, we define $f^{\#}: R^n \to [0,\infty]$ by

$$f^{\#}(x) = \sup_{x \in Q} m_Q |f - m_Q f| .$$

Hence, $f^{\#} \in L^\infty$ if and only if $f \in BMO$, and $f^{\#} \leq 2f^*$.

For each of the variations of the definition of the maximal function we have given, there corresponds a variation of the #-function. In particular, $f^{\#}_\delta$ will denote the dyadic #-function.

The basic property of the #-function is the following:

Theorem: Let $\omega \in A_\infty$, $1 < p < \infty$, and $f \in L^1_{loc}(R^n)$. If $\inf(1, f^*) \in L^p(\omega dx)$, then

$$\int_{R^n} (f^*_\delta)^p \omega dx \leq C(p,\omega) \int_{R^n} (f^{\#}_\delta)^p \omega dx .$$

Note that the same inequality is true for the non-dyadic case. Since we are mainly interested in the case where $\omega \in A_p$, what we really need is

$$\int_{R^n} |f|^p \omega dx \leq C(p,\omega) \int_{R^n} (f^{\#})^p \omega dx ,$$

which follows from the theorem, and from the facts that $|f| \leq f^*_\delta$ a.e. and $f^{\#}_\delta \leq f^{\#}$.

The proof of the theorem is an example of how the good λ's inequality (see section II of chapter 0) is useful for proving weighted norm inequalities.

We wish to prove that for $\omega \in A_\infty$ there exists $\gamma = \gamma(\omega, p) > 0$ such that

(1) $\omega(\{f^*_\delta > 2\lambda , f^{\#}_\delta < \gamma\lambda\}) \leq \nu_p \omega(\{f^*_\delta > \lambda\})$ where ν_p satisfies $2^p \nu_p < 1$. In fact, we shall prove

(2) $|\{f^*_\delta > 2\lambda, f^{\#}_\delta < \gamma\lambda\}| \leq 2^n \gamma |\{f^*_\delta > \lambda\}|$ for all $\gamma > 0$. Because the proof will be effected through a decomposition of $\{f^*_\delta > \lambda\}$ into maximal disjoint (dyadic) cubes, the characterization of A_∞ weights (see the second lemma of section II in chapter 2) implies that (1) follows from (2), for γ small enough.

Let $\Omega_\lambda = \{f^*_\delta > \lambda\}$, and assume for the moment that $f \in L^p(\omega dx)$. Then, from $\omega \in A_\infty$ we obtain (e.g., apply the aforementioned lemma) $\omega(\tilde{Q}) \geq C(\omega)\omega(Q)$ for any dyadic cube Q, where $C(\omega) > 1$. These two facts imply that Ω_λ cannot contain an infinite increasing sequence of dyadic cubes, and hence we may consider the decomposition of

Ω_λ into disjoint maximal dyadic cubes contained in it.

It is enough to prove that for each such cube Q, and for every $\gamma > 0$.

(2′) $|\{x \in Q : f_\delta^* > 2\lambda\ ,\ f_\delta^\# < \gamma\lambda\}| \leqq 2^n \gamma |Q|$. Since Q is maximal, each dyadic cube Q' containing Q satisfies $m_{Q'}|f| \leqq \lambda$. Hence, for $x \in Q$, $f_\delta^*(x) > 2\lambda$ implies that $(f\chi_Q)_\delta^*(x) > 2\lambda$, and, since $m_{\tilde{Q}}|f| < \lambda$, we obtain that $x \in Q$ and $f_\delta^*(x) > 2\lambda$ imply $((f - m_{\tilde{Q}} f)\chi_Q)_\delta^* > \lambda$.

From the weak type (1,1) estimate we now obtain

$$|\{((f - m_{\tilde{Q}} f)\chi_Q)_\delta^* > \lambda\}| \leqq \frac{1}{\lambda} \int_Q |f - m_{\tilde{Q}} f| dx$$

$$\leqq \frac{2^n |Q|}{\lambda} \frac{1}{|\tilde{Q}|} \int_{\tilde{Q}} |f - m_{\tilde{Q}} f| dx$$

$$\leqq \frac{2^n |Q|}{\lambda} \inf_{x \in Q} f_\delta^\#(x) \quad .$$

Hence, (2′) follows immediately if $\inf_{x \in Q} f_\delta^\#(x) < \gamma\lambda$, while, if not, (2′) holds trivially.

To get rid of the assumption that $f \in L^p(\omega dx)$, we consider $f_n = \begin{cases} f, & |f| \leqq n \\ n \operatorname{sgn} f, & |f| > n \end{cases}$ which does lie in $L^p(\omega, dx)$, by assumption. Then we can obtain the appropriate norm inequality for f_n , and take the limit as n tends to infinity. This completes the proof.

We should remark that we have really proved a bit more. In particular, the preceeding calculation shows that for all γ, λ , and μ positive,

(3) $$|\{x \in Q : f_\delta^*(x) > \lambda + \mu\ ,\ f_\delta^\# < \gamma\mu\}|$$

$$\leqq 2^n \gamma |Q|$$

for each maximal dyadic cube Q in Ω_λ . This implies that $(f_\delta^*)_\delta^\# \leqq C(f_\delta^\#)_\delta^*$.

Indeed, let Q_0 be any given dyadic cube, and let $\lambda_0 > \inf_{Q_0} f_\delta^*$. Consider the maximal dyadic cubes in Ω_λ for $\lambda = \lambda_0 + \mu$. Then Q_0 is the union of all such cubes that it intersects. (This follows from the fact that any two dyadic cubes are either disjoint or one is contained in the other, and the fact that Q_0 is not contained in

Ω_λ , by our choices of λ and λ_0 .) If we apply (3) above and sum over all such maximal dyadic cubes that also happen to be included in Q_0 , then we obtain that

$$|\{x \in Q_0 : (f_\delta^*(x) - \lambda_0) > 2\mu \ , \ f_\delta^\#(x) < \gamma\mu\}|$$

$$\leq 2^n\gamma|\{x \in Q_0 : (f_\delta^*(x) - \lambda_0) > \mu\}| \quad .$$

For $\gamma > 0$ small enough, we may apply the good λ's inequality (which is still true when $(\mathbb{R}^n, dx)$ is replaced by (Q_0, dx) , or any measure space, for that matter) to conclude that

$$\int_{Q_0} \max(f_\delta^* - \lambda_0, 0)dx \leq C \int_{Q_0} f_\delta^\# \, dx \ ,$$

with C independent of λ_0 . If we let λ_0 tend to $\inf_{x \in Q} f_\delta^*(x)$, then we obtain

$$\int_{Q_0} \left(f_\delta^* - \inf_{y \in Q_0} f_\delta^*(y)\right)dx \leq C \int_{Q_0} f_\delta^\# dx \ ,$$

which is clearly stronger than $(f_\delta^*)_\delta^\# \leq C(f_\delta^\#)_\delta^*$.

This inequality also holds for the non-dyadic versions of these operators.

We now apply the # function to an interpolation problem.

Theorem: <u>Let</u> T <u>be a sublinear operator which is bounded from</u> L_c^∞ <u>to</u> BMO <u>and from</u> $H^{1,\infty}$ <u>to</u> L^1 . <u>Then</u> T <u>extends boundedly to every</u> $L^p(\mathbb{R}^n, dx)$, $1 < p < \infty$.

<u>Proof</u>: Observe that the functions in L_c^∞ with mean zero are dense in every L^p , $1 < p < \infty$, and hence it is enough to prove $\|Tf\|_p \leq C\|f\|_p$ (with C independent of f) for such functions. We denote this function space by L_0^∞ .

In fact we shall show that $\|(Tf)^\#\|_p \leq C\|f\|_p$ and apply the previous theorem. To be able to do this, we must show that $\inf(1, (Tf)^*) \in L^p(dx)$ for $f \in L_c^\infty$ with mean zero. But these functions lie in $H^{1,\infty}$, so that $Tf \in L^1$ and $(Tf)^* \in$ weak-L^1 ; hence, $\inf(1, (Tf)^*) \in L^p$ for $p > 1$.

We are going to prove estimates of the form

$$|\{(Tf)^\# > \lambda\}| \leq \frac{C\|f\|_p^p}{\lambda^p}$$

for all $\lambda > 0$ and any given p , $1 < p < \infty$. Thus, by the Marcinkiewicz interpolation theorem the desired strong type estimates also hold.

Suppose $f \in L_c^\infty$ has mean 0 and that $\lambda > 0$. Let $\{Q_j\}$ be the collection of cubes associated with the Calderón-Zygmund decomposition of $|f|^p$ corresponding to the value λ^p. We write

$$f = b + g = \sum_{j=1}^{\infty} (f - m_{Q_j} f)\chi_{Q_j} + g .$$

We then have:

$$m_{Q_j}(|f|^p) \leq 2^n \lambda^p$$

$$\|g\|_\infty \leq 2^{\frac{n}{p}} \lambda .$$

Let $C_0 > 0$. Then

$$\{(Tf)^{\#} > (C_0 + 1)\lambda\} \subseteq \{(Tg)^{\#} > C_0\lambda\} \cup \{(Tb)^{\#} > \lambda\} .$$

Since T is bounded from L_c^∞ to BMO, we can choose C_0 big enough so that $\{(Tg)^{\#} > C_0\lambda\} = \emptyset$.

To estimate $\{(Tb)^{\#} > \lambda\}$, we proceed as follows. For $j \in \mathbb{N}$ let $f_j = (f - m_{Q_j} f)\chi_{Q_j}$, so that

$$\Big[\int_{Q_j} |f_j|^p dx\Big]^{\frac{1}{p}} \leq 2|Q_j|^{\frac{1}{p}} (m_{Q_j} |f|^p)^{\frac{1}{p}}$$

$$\leq C\lambda\, |Q_j|^{\frac{1}{p}} .$$

Hence, $\|f_j\|_{H^{1,p}} \leq C\lambda\, |Q_j|$, so that $\|b\|_{H^{1,p}} \leq C\lambda\, |\Omega_\lambda|$. (Here Ω_λ is $\{x : (|f|^p)^*(x) > \lambda^p\}$, as in the construction of the Calderon-Zygmund decomposition.) By the weak type (1,1) estimate for the maximal function $|\Omega_\lambda| \leq \dfrac{\|f\|_p^p}{\lambda^p}$, and so we obtain

$$|\{(Tb)^{\#} > \lambda\}| \leq |\{(Tb)^* > \tfrac{\lambda}{2}\}|$$

$$\leq \frac{C\|Tb\|_1}{\lambda}$$

$$\leq \frac{C\|b\|_{H^{1,p}}}{\lambda} \leq C|\Omega_\lambda| \leq C\frac{\|f\|_p^p}{\lambda^p} \quad .$$

This completes the proof of our interpolation theorem. Note that our arguments may be extended to the case where the functions take values in a Banach space, since the identification of $H^{1,\infty}$ and $H^{1,p}$, $1<p<\infty$, remains true in this context, as we pointed out in the last section.

CHAPTER 4

Calderón-Zygmund Operators

I. Introduction.

The original notion of a Calderón-Zygmund operator was introduced by Calderón and Zygmund in [CZ]. Their main object was to generalize the Hilbert transform and similar operators associated with $\mathbb{R}$ to higher dimensions. Motivations for these generalizations can be found in [F].

The original Calderón-Zygmund operators were convolution operators defined by singular kernels:

$$Tf(x) = \lim_{\varepsilon \to 0} \int_{\|x-y\|>\varepsilon} \Omega\left(\frac{x-y}{\|x-y\|}\right) \frac{1}{\|x-y\|^n} f(y)dy \ .$$

Here, Ω is a function defined on the sphere S^{n-1}, satisfying suitable conditions and, in particular, $\int_{S^{n-1}} \Omega(x)d\sigma = 0$ (where $d\sigma$ denotes the Lebesgue surface measure on S^{n-1}). Under these assumptions, it turns out that the above limit exists a.e. if f is in some $L^p(\mathbb{R}^n, dx)$ for $p \in [1, +\infty[$.

To study the L^p-boundedness of such operators, one begins by studying the case $p = 2$, which can be handled via Plancherel's theorem. The case $p \neq 2$ follows from the case $p = 2$ and from the properties of the kernel k, defined by

$$k(x,y) = \Omega\left(\frac{x-y}{\|x-y\|}\right) \frac{1}{\|x-y\|^n} \ .$$

Thus, k is defined on the complement of the diagonal, $\Delta = \{(x,y) \in \mathbb{R}^n \times \mathbb{R}^n / x = y\}$, in $\mathbb{R}^n \times \mathbb{R}^n$.

Note that the Hilbert transform corresponds to $\Omega(x) = x$ on $S_0 = \{-1, 1\}$. The Riesz transforms, denoted by R_j, $j \in [1,n]$, in several senses the most natural generalization of the Hilbert transform to several dimensions, correspond to $\Omega_j(x) = x_j$, the j-th coordinate of x .

Having noticed that the second part of the general plan sketched above could be carried out regardless of whether k was a convolution kernel or not, Coifman and Meyer introduced in [CM2] a new notion of Calderón-Zygmund operators, which has the advantage of containing several classes of operators which are not convolution operators (for examples see Chapters V and VII).

Moreover, these new Calderón-Zygmund operators are, like the Hilbert transform and the Riesz transforms, bounded on weighted L^p's , the weight being in A_p .

The point is to assume the L^2-boundedness of the operator T and to impose on the kernel K exactly the size and smoothness assumptions which permit us to go from L^2 to $L^p (p \neq 2)$. It can be seen very easily that these are exactly the assumptions satisfied by the kernels of the form $K(x,y) = \Omega\left(\frac{x-y}{\|x-y\|}\right) \cdot \frac{1}{\|x-y\|^n}$ when Ω is continuous on S^{n-1} and has distributional derivatives of order one that are essentially bounded.

<u>Definition 1</u>: <u>A standard kernel is a continuous function</u> $K : \Delta^c \to C$ <u>for which there exists a positive constant</u> C <u>such that for all</u> $(x,y) \in \Delta^c$,

$$K(x,y) \leqq \frac{C}{|x-y|^n}$$

and

$$|\nabla_x K(x,y)| + |\nabla_y K(x,y)| \leqq \frac{C}{|x-y|^{n+1}} .$$

<u>The gradients are taken in the distributional sense, and they are assumed to be functions</u>.

The smallest constant for which these inequalities hold is denoted $C(K)$, and is called the constant of the kernel K . We shall often refer to the above inequalities as the <u>standard estimates</u> (on the kernel K).

Some immediate consequences of the standard estimates are the following.

(i) For $Q \subseteqq R^n$, $x \in Q$, $f \in L^1_{loc}(R^n)$,

$$\int_{\bar{Q} \setminus Q} |K(x,y)| \; |f(y)| dy \leqq Cf^*(x) .$$

(ii) For $Q \subseteqq R^n$ and $x, x_0 \in Q$,

$$\int_{\bar{Q}^c} |K(x,y) - K(x_0,y)| \; |f(y)| \, dy \leqq Cf^*(x_0) .$$

(iii) For any $Q \subseteqq R^n$ and $y_0 \in Q$,

$$\int_{\bar{Q}^c} \int_Q |K(x,y) - K(x,y_0)| \; |f(y)| \, dydx \leqq C|Q| f^*(y_0) .$$

Moreover, we shall see that these conditions are the only properties of K that we shall use.

Definition 2: An operator T taking $C_c^\infty(\mathbb{R}^n)$ into $L^1_{loc}(\mathbb{R}^n, dx)$ is a Calderón-Zygmund operator (CZO) iff:

(i) T extends to a bounded linear operator on $L^2(\mathbb{R}^n, dx)$;

(ii) there exists a standard kernel K such that for every $f \in L_c^\infty(\mathbb{R}^n)$,

$$Tf(x) = \int K(x,y) f(y) dy \quad \text{a.e.}$$

on $\{\text{supp}\, f\}^c$.

Note that if (i) holds, then it is enough to check (ii) for some subclass of $L_c^\infty(\mathbb{R}^n)$ which is dense in $L^2(\mathbb{R}^n, dx)$, e.g., $C_c^\infty(\mathbb{R}^n)$.

If T is a CZO , then we let $\|T\|_{CZ} = \|T\|_{2,2} + C(K)$.

There is an important extension of the notion of CZO , which we shall use throughout chapter 6. Let (B_1, B_2) be a pair of Banach spaces, and let $\mathcal{B}(B_1, B_2)$ be the Banach space of bounded linear transformations of B_1 into B_2 . Then a (B_1, B_2)CZO is an operator T which sends C^∞_{c,B_1} into $L^1_{loc,B_2}(\mathbb{R}^n, dx)$, which is associated to a kernel taking values in $\mathcal{B}(B_1, B_2)$, and where the appropriate analogues of the above properties of T and K still hold. We emphasize the fact that the arguments that we shall use to treat the case of $B_1 = B_2 = \mathbb{C}$ will have obvious extensions to the general case.

The most basic examples of CZO's are given by the convolution kernels. There exists a simple necessary condition for a standard convolution kernel to be the kernel of a CZO , and it is also sufficient if B_1 and B_2 are Hilbert spaces. The condition is

$$\sup_{0<\epsilon<\eta<\infty} \left| \int_{\epsilon<|x|<\eta} K(x,0) dx \right| < \infty .$$

See [S] for the sufficiency of this condition, and see p. 67 for the necessity.

II. Action of a CZO on $H^{1,\infty}$ and L_0^∞ .

In order to appreciate the role of the standard estimates, let us see how they imply that a Calderón-Zymund operator is bounded from $H^{1,\infty}$ to L^1 , and from L_0^∞ to BMO . (Hence, T is bounded on all $L^p(\mathbb{R}^n, dx)$, $1<p<\infty$, by the interpolation theorem at the end of the preceeding chapter.) The property of T that we shall really use is weaker than L^2 boundedness (but, as we shall see, it is equivalent to L^2-boundedness for a CZO). Let us call this property (H) :

For $Q \subseteq R^n$, $a \in L^\infty(R^n)$ such that $\operatorname{supp} a \subseteq Q$, then

(H)

$$\int_{\bar{Q}} |Ta|\,dx \leq C\|a\|_\infty |Q| \quad .$$

If T is bounded on $L^2(R^n, dx)$, then T satisfies (H), since

$$\int_{\bar{Q}} |Ta|\,dx \leq |\bar{Q}|^{\frac{1}{2}} \Big(\int_{\bar{Q}} |Ta|^2 dx\Big)^{\frac{1}{2}}$$

$$\leq C|\bar{Q}|^{\frac{1}{2}} \Big(\int_{Q} |a|^2 dx\Big)^{\frac{1}{2}}$$

$$\leq C|Q|\ \|a\|_\infty \quad .$$

That the converse holds is a consequence of the following theorem and the interpolation result at the end of the last chapter.

Theorem: Let T be associated to a standard kernel. Then the following are equivalent.

(i) T satisfies H.

(ii) T is a bounded map from $H^{1,\infty}$ into L^1.

(iii) T is a bounded map from L^∞_0 to BMO.

Proof: (i) implies (iii).

Let $a \in L^\infty_c$ and Q a cube with center x_0. Put $a_1 = a\chi_{\bar{Q}}$ and $a_2 = a - a_1$. From (H) we obtain that

$$\frac{1}{|Q|} \int_Q |Ta_1|\,dx \leq C\|a\|_\infty \ .$$

Thus, to estimate $\|Ta\|_*$, it is enough to show that

$$\frac{1}{|Q|} \int_Q |Ta_2 - Ta_2(x_0)|\,dx \leq C\|a\|_\infty \ .$$

To do this we use the standard estimates:

$$\frac{1}{|Q|}\int_Q |Ta_2 - Ta_2(x_0)|dx$$

$$\leq \frac{1}{|Q|}\int_Q \int_{(\bar{Q})^c} |K(x,y) - K(x_0,y)|\ |a(y)|\,dydx$$

$$\leq C\|a\|_\infty .$$

(iii) implies (ii).

Let Q be a cube and let $a \in \mathcal{a}_Q^\infty$, so that we want to show $\|Ta\|_1 \leq C$.

$$\int_{\bar{Q}^c} |Ta|dx = \int_{\bar{Q}^c} |\int_Q K(x,y)a(y)dy|dx \quad .$$

Because $m_Q a = 0$, it follows that, for any y_0 in Q,

$$\int_{(\bar{Q})^c} |Ta|dx = \int_{\bar{Q}^c} |\int_Q (K(x,y) - K(x,y_0))a(y)dy|dx$$

$$\leq C|Q|\ \|a\|_\infty \leq C .$$

Let Q' be a cube of the same size as $\bar{Q}$, and such that Q' and Q have a face in common but have disjoint interiors. (That is, they are adjacent cubes.) By (iii), $\|Ta\|_* \leq \frac{C}{|Q|}$, and by the above calculation, $|m_{Q'}Ta| \leq \frac{C}{|Q|}$. Now, a standard BMO calculation shows that

$$|m_{\bar{Q}}Ta - m_{Q'}Ta| \leq C\|Ta\|_* \leq \frac{C}{|Q|} ;$$

hence

$$\frac{1}{|\bar{Q}|}\int_{\bar{Q}} |Ta|\,dx$$

$$\leq \frac{1}{|\bar{Q}|}\int_{\bar{Q}} |Ta - m_{\bar{Q}}Ta|\,dx + |m_{\bar{Q}}Ta - m_{Q'}Ta| + |m_{Q'}Ta|$$

$$\leq C\|Ta\|_* + \frac{C}{|Q|} + \frac{C}{|Q|} \leq \frac{C}{|Q|} \quad .$$

Putting together these two calculations, we obtain $\|Ta\|_1 \leq C$, and (ii) is an immediate consequence.

(ii) implies (i).

Let Q be any cube, $a \in L^\infty$, $\operatorname{supp} a \subseteq Q$, and let Q' be a cube of the same size as Q , with the property that $\bar{Q}$ and $\bar{Q}'$ have disjoint interiors and a face in common. Choose $\tilde{a} \in L^\infty(\mathbb{R}^n)$ so that $\operatorname{supp} \tilde{a} \subseteq Q \cup Q'$, $\tilde{a} = a$ on Q , $\|\tilde{a}\|_\infty = \|a\|_\infty$, and $\int_{\mathbb{R}^n} \tilde{a}\, dx = 0$, and define a' by $\tilde{a} = a + a'$. Then $\|\tilde{a}\|_{H^{1,\infty}} \leq C\|a\|_\infty |Q|$, and, therefore,

$$\int_{\bar{Q}} |T\tilde{a}|\, dx \leq C\|a\|_\infty |Q| \quad .$$

On the other hand, the first of the standard estimates on K implies that

$$|Ta'| \leq C\|a\|_\infty$$

on $\bar{Q}$, so that

$$\int_{\bar{Q}} |Ta|\, dx \leq C\|a\|_\infty |Q| \ ,$$

which is (i). Hence, the theorem is proved.

Observe that any operator T which is bounded on $L^p(\omega dx)$ for $1<p<\infty$ and some $\omega \in A_p$ satisfies (H) above. Indeed, let $a \in L^\infty(\mathbb{R}^n)$ and suppose that $\operatorname{supp} a$ is contained in a cube Q . Then, by Hölder's inequality and the A_p condition on ω ,

$$\frac{1}{|Q|} \int_{\bar{Q}} |Ta|\, dx \leq C \left[\frac{\int_Q |Ta|^p \omega\, dx}{\int_Q \omega\, dx} \right]^{\frac{1}{p}}$$

$$\leq C \left[\frac{\int_Q |a|^p \omega\, dx}{\int_Q \omega\, dx} \right]^{\frac{1}{p}}$$

$$\leq C\|a\|_\infty \quad .$$

Thus, by the theorem and the remarks preceding it, if T is a linear operator associated to a standard kernel which is bounded on $L^p(\omega dx)$ for some $\omega \in A_p$, $1<p<\infty$, then T is CZO . The converse is also true, and this is the topic that we consider next.

III. Action of CZO's on a Weighted L^p Space.

Theorem: If $1<p<\infty$ and $\omega \in A_p$, then every CZO T is bounded on $L^p(\omega dx)$.

The following converse is also true. (See [CF].) If a weight ω is such that the Hilbert transform (in the one dimensional case) or any Riesz transform (in the multi-dimensional case), is bounded on $L^p(\mathbb{R}^n,\omega dx)$, then $\omega \in A_p$. Note that this characterization of A_p-weights is a reason for their interest among harmonic analysts.

The Fefferman-Stein Inequality:

Let T be a CZO and $p>1$. Then for $f \in L_c^\infty(\mathbb{R}^n)$,

$$(Tf)^{\#} \leq C(p,T)((|f|^p)^*)^{\frac{1}{p}} .$$

To prove this, let $f \in L_c^\infty$ and let Q be a cube with center x_0. Put $f_1 = f\chi_{\overline{Q}}$ and $f_2 = f - f_1$. Then (using the standard estimates),

$$\frac{1}{|Q|}\int_Q |Tf - Tf_2(x_0)|dx \leq \frac{1}{|Q|}\int_Q |Tf_1(x)|dx + \frac{1}{|Q|}\int_Q |Tf_2(x) - Tf_2(x_0)|dx$$

$$\leq \left[\frac{1}{|Q|}\int_Q |Tf_1|^p dx\right]^{\frac{1}{p}} + cf^*(x_0)$$

$$\leq C\left[\frac{1}{|Q|}\int_{\overline{Q}} |f_1|^p dx\right]^{\frac{1}{p}} + cf^*(x_0)$$

$$\leq C((|f|^p)^*)^{\frac{1}{p}}(x_0) ,$$

which implies the result. Note that the second inequality was obtained using the standard estimates on the kernel K in the usual way, and the third inequality comes from the boundedness of T on $L^p(\mathbb{R}^n, dx)$ (which we observed in the previous section).

Suppose now that $f \in L_c^\infty$ and $\omega \in A_p$, so that we wish to show

$$\|Tf\|_{L^p(\omega dx)} \leq C\|f\|_{L^p(\omega dx)} .$$

To apply the Fefferman-Stein inequality, we need

$$\|(Tf)\|_{L^p(\omega dx)} \leqq C\|Tf^{\#}\|_{L^p(\omega dx)} .$$

We know that this is true as long as we have the following control on $(Tf)^*$ (see Section III of the last chapter):

<u>Lemma</u>: Let $\omega \in A_p$ and $f \in L_c^\infty$. Then

$$\inf(1, (Tf)^*) \in L^p(\omega dx) .$$

Assuming the lemma for the moment, let us finish the proof of the theorem. Since $\omega \in A_p$, there exists $r = r(\omega) > 1$ such that $\omega \in A_{\frac{p}{r}}$. (See Section II of Chapter 2.) Applying the Fefferman-Stein inequality, we obtain (for $f \in L_c^\infty(\mathbb{R}^n)$)

$$\|Tf\|_{L^p(\omega dx)} \leqq \|(Tf)^{\#}\|_{L^p(\omega dx)}$$

$$\leqq C\|((|f|^r)^*)^{\frac{1}{r}}\|_{L^p(\omega dx)}$$

$$\leqq C\|(|f|^r)^*\|^{\frac{1}{r}}_{L^{\frac{p}{r}}(\omega dx)}$$

$$\leqq C\||f|^r\|^{\frac{1}{r}}_{L^{\frac{p}{r}}(\omega dx)}$$

$$\leqq C\|f\|_{L^p(\omega dx)} ,$$

which implies the theorem.

To prove the lemma, note that if $f \in L_c^\infty$, then Tf lies in $L^1_{loc}(\mathbb{R}^n, dx)$ and is dominated by $C\frac{1}{|x|^n}$ at infinity. Thus, $(Tf)^*(x)$ is dominated by $C\frac{\log|x|}{|x|^n}$ at infinity.

Let Q be the unit cube. Then $((\chi_Q)^*)^*$ grows like $C\frac{\log|x|}{|x|^n}$ as x increases. Hence,

$$\int_{\mathbb{R}^n} |\inf(1, (Tf)^*)|^p \omega dx \leqq C + C\int_{\mathbb{R}^n} |(\chi_Q^*)^*|^p \omega dx < \infty ,$$

since $\omega \in A_p$; the lemma now follows.

For $p = 1$, there is a weak type result which is analogous to the above theorem, which is as follows:

__Theorem__: __If__ $\omega \in A_1$, __then every__ CZO __takes__ $L^1(\omega dx)$ __into__ weak-$L^1(\omega dx)$ __boundedly__.

See [CM2] for a proof of this when $\omega \equiv 1$. The general case is proved in the same way, making minor modifications.

Originally, this weak type result was used to prove L^p boundedness for $1 < p < 2$, by interpolation, and then the case $p > 2$ was obtained through a duality argument. This path is not available to us, since we want only to use arguments which extend to the case where the functions take their values in a Banach space, in which case it is not always true that

$$(L^p_B(R^n, dx))^* = L^{p'}_{B^*}(R^n, dx) ,$$

where $\frac{1}{p} + \frac{1}{p'} = 1$.

IV. __Singular Integral Operators__.

So far, we have dealt with operators "associated", in a precise sense, to a standard kernel, but we have not considered the question of whether a CZO is determined by its kernel. The answer is no; the identity is a CZO with kernel identically zero. In fact, the following proposition (whose proof is a simple exercise in measure theory) shows that a CZO is determined by its kernel, up to adding a multiplication operator:

__Proposition__: __Suppose that__ T __is a bounded operator on__ L^2 , __and that if__ $f \in L^\infty_c$, __then__ $Tf = 0$ a.e. __on__ $\{\text{supp}\, f\}^c$. __Then, for all__ $f \in L^2$, $Tf = bf$, __where__ $b \in L^\infty$.

Since most of the classical convolution operators are defined by a principal value integral, it is natural to ask which conditions on K will ensure that the principal value defined by K is meaningful. A necessary condition is easy to find.

Suppose that, for each $f \in C^\infty_c(R^n)$,

$$\lim_{\epsilon \to 0} \int_{|x-y|>\epsilon} K(x,y) f(y)\, dy$$

exists a.e. Then, since for such f the integrals

$$\int_{|x-y|>1} |K(x,y)|\, |f(y)|\, dy$$

are uniformly bounded, choosing f constant near x, one sees that the a.e. existence of

$$\lim_{\epsilon\to 0} \int_{1>|x-y|>\epsilon} K(x,y)dy$$

is a necessary condition.

Conversely, if K satisfies this condition, then the above principal value converges a.e. for all $f \in C_c^\infty(R^n)$. If K is the kernel of a CZO, then this a.e. convergence of the principal value occurs for more general functions, as the following theorem shows. In this case, K is called a CZK (Calderón-Zygmund kernel), and its "norm" is defined to be the CZ norm of the corresponding CZO.

<u>Theorem</u>: <u>Let</u> K <u>be a Calderón-Zygmund kernel</u>, $1 \leqq p < \infty$, $\omega \in A_p$, <u>and</u> $f \in L^p(\omega dx)$. <u>Then</u>

$$\lim_{\epsilon\to 0} \int_{\epsilon<|x-y|} K(x,y)f(y)dy$$

<u>exists</u> a.e.

Let us first show that this result is local. To see this we show that, for a fixed $\epsilon > 0$, the above integral converges absolutely. Indeed, if $p = 1$ and $\omega \equiv 1$, then this is obvious. If $p = 1$ but $\omega \not\equiv 1$, then we need to show that $\frac{1}{|x-y|^n \omega(y)}$ is bounded, for each x, on $\{y : |x-y| > \epsilon\}$. This follows from $\omega(y) \geqq C\omega^*(y) \geqq C(x,\epsilon)\frac{1}{|x-y|^n}$.

If $\infty > p > 1$, then $\omega^{-\frac{1}{p-1}} \in A_{p'}, (\frac{1}{p} + \frac{1}{p'} = 1)$; hence $(\chi_Q)^* \in L^{p'}(\omega^{-\frac{1}{p-1}}dx)$. Let Q be, say, the unit cube. But, then, for x fixed, $(\chi_Q)^*(y)$ decreases like $\frac{1}{|x-y|^n}$ for y large, and hence

$$\inf\left(\frac{1}{|x-y|^n}, 1\right) \in L^{p'}(\omega^{-\frac{1}{p-1}}dx) .$$

Hence, by Hölder's inequality,

$$\int_{|x-y|>\epsilon} |K(x,y)|\,|f(y)|dy$$

$$\leqq \left(\int_{|x-y|>\epsilon} \left(\frac{c}{|x-y|^n}\right)^{p'} \omega^{-\frac{1}{p-1}} dy\right)^{1-\frac{1}{p}} \left(\int_{|x-y|>\epsilon} |f(y)|^p \omega dy\right)^{\frac{1}{p}} < \infty .$$

These calculations, therefore, show that the conclusion of the theorem is of a local nature. Since $L^p(\omega dx) \subseteq L^1_{loc}(dx)$ for $\omega \in A_p$, it is enough to prove the result for $p = 1$. Because we already know, by hypothesis, that the theorem holds for a dense class of functions (namely, $C_c^\infty(R^n)$), it is enough to prove a maximal inequality.

For any CZO T, consider the following operators:

$$T_\epsilon f(x) = \int_{|x-y|>\epsilon} K(x,y) f(y) dy$$

$$T_* f(x) = \sup_{\epsilon > 0} |T_\epsilon f(x)| .$$

(For the sake of consistency, we take the norm $|x|$ for $x \in R^n$ to be one whose unit ball is a cube.) Though it can be shown that the class of singular integrals depends on the choice of the norm on R^n (see [CM2]), the change of norm is not relevant for the problem considered here. This is a consequence of the first standard estimate.

The aforementioned maximal inequality, and, hence, the theorem, is a consequence of the following:

<u>Proposition</u>: <u>If</u> T <u>is a</u> CZO, <u>then</u> T_* <u>sends</u> $L^1(\omega dx)$ <u>boundedly into weak-</u> $L^1(\omega dx)$ <u>for</u> $\omega \in A_1$.

This proposition follows immediately from the next result, which also shows that T_* is bounded on $L^p(\omega dx)$ for $\omega \in A_p$ and $1 < p < \infty$.

<u>Cotlar's Inequality</u>:

<u>Let</u> $0 < \delta \leq 1$ <u>and</u> T <u>be a</u> CZO. <u>Then there exists a</u> $C_\delta > 0$ <u>such that if</u> $f \in L^p(\omega dx)$, <u>for some</u> $\omega \in A_p$ <u>and</u> $1 \leq p < \infty$, <u>then</u>

$$T_* f(x) \leq C_\delta [([(Tf)^\delta]^*)^{\frac{1}{\delta}} + \|T\|_{CZ} f^*(x)] .$$

To prove this result, we begin with the case $\delta = 1$. Then it is enough to show that for all $\epsilon > 0$,

$$|T_\epsilon f(0)| \leq C_n [(Tf)^*(0) + \|T\|_{CZ} f^*(0)] .$$

Let $Q = \{y \in R^n : |y| < \frac{\epsilon}{2}\}$. Take $f \in L^1_{loc}$ and set $f_1 = f\chi_{\overline{Q}}$ and $f_2 = f - f_1$, so that

$$T(f_2)(0) = T_\epsilon f(0)$$

and, for $x \in Q$,

$$|T(f_2)(0) - Tf_2(x)| \leq c\|T\|_{CZ} f^*(0) .$$

(This latter inequality is obtained by the usual calculation involving the standard estimates on K). Hence,

$$|T_\epsilon f(0)| \leq |Tf(x)| + |Tf_1(x)| + c_n f^*(0)\|T\|_{CZ} .$$

If $T_\epsilon f(0) = 0$, then there is nothing to prove. Otherwise, choose $0 < \lambda < |T_\epsilon f(0)|$, Then for each $x \in Q$, either $|Tf(x)| > \frac{\lambda}{3}$, or $|Tf_1(x)| > \frac{\lambda}{3}$, or $cf^*(0)\|T\|_{CZ} > \frac{\lambda}{3}$. That is, either

$$\lambda < 3cf^*(0)\|T\|_{CZ} ,$$

or

$$Q = \{x \in Q : |Tf(x)| > \tfrac{\lambda}{3}\} \cup \{x \in Q : |Tf_1(x)| > \tfrac{\lambda}{3}\} .$$

But,

$$|\{x \in Q : |Tf(x)| > \tfrac{\lambda}{3}\}| \leq \frac{3|Q|}{\lambda}(Tf)^*(0)$$

and

$$|\{x \in Q : |Tf_1(x)| > \tfrac{\lambda}{3}\}| \leq \frac{c}{\lambda}\|f_1\|_{L^1}\|T\|_{CZ}$$

$$\leq \frac{c|Q|}{\lambda} f^*(0)\|T\|_{CZ} .$$

(For the second inequality we use the weak type $(1,1)$ estimate on T.)

In both cases we obtain

$$\lambda < 3(Tf)^*(0) + cf^*(0)\|T\|_{CZ} ,$$

and the desired inequality follows.

In the case where $0 < \delta < 1$, we have

$$|T_\epsilon f(0)|^\delta \leq c_\delta [|Tf(x)|^\delta + |Tf_1(x)|^\delta + (c\|T\|_{CZ} f^*(0))^\delta] .$$

If we take the mean over Q of both sides, and then raise the resulting expressions to the $\frac{1}{\delta}$th power, we obtain

$$|T_\varepsilon f(0)| \leqq C_\delta ((|Tf|^\delta)^*)^{\frac{1}{\delta}}(0)$$

$$+ C_\delta \|T\|_{CZ} f^*(0)$$

$$+ C_\delta \frac{1}{|Q|}\int_Q |Tf_1(x)|^\delta dx]^{\frac{1}{\delta}} .$$

We estimate this latter term by using the weak-L^1 boundedness of T and Kolmogorov's inequality (see Section II, Chapter 0).

$$\frac{1}{|Q|}\int_Q |Tf_1(x)|^\delta dx \leq C_\delta |Q|^{-\delta} \|f_1\|_1^\delta \|T\|_{CZ}^\delta$$

$$\leqq C_\delta \|T\|_{CZ}^\delta (f^*(0))^\delta .$$

Hence Cotlar's inequality follows.

V. Some Applications of the Boundedness of T_* .

The operator T_* has some advantages over the Calderón-Zygmund operator T . For example, T_*f is lower semicontinuous, and that makes T_* a good operator to deal with when one uses the good λ's inequality. (See p 6 .) On the other hand, T_* is the supremum of nonsingular operators, which are easier to manipulate than T . Hence the boundedness of T_* will often be used to pass from the T_ε's to T via the dominated convergence theorem. Let us consider first an example of this where the T_ε's approximate T , in the sense that

$$Tf(x) = \lim_{\varepsilon \to 0} T_\varepsilon f(x) \quad \text{a.e.}$$

(In this case, we say that T is a Calderón-Zygmund singular integral, or CZSI . Note that the kernel associated to T is a CZK .)

<u>Proposition</u>: <u>Let</u> (A,μ) <u>be a measure space, and for each</u> $u \in A$, <u>let</u> K_u <u>be a standard kernel defining a</u> CZSI T_u . <u>Suppose that</u>

(i) $\int_A \|T_u\|_{CZ} d\mu < \infty$

and

(ii) $(x,y,u) \to K_u(x,y)$ is measurable on $\Delta^c \times A$.

Then $K = \int_A K_u \, d\mu$ determines a CZSI T .

Proof: K automatically satisfies the standard estimates, and

$$C(K) \leqq \int_A C(K_u) d\mu \leqq \int_A \|T_u\|_{CZ} d\mu \ .$$

For $\epsilon > 0$, let us write

$$K_\epsilon(x,y) = K(x,y) \chi_{\{|x-y| > \epsilon\}} \ .$$

Thus, we must show that for $f \in C_c^\infty(R^n)$,

$$\lim_{\epsilon \to 0} \int_{R^n} K_\epsilon(x,y) f(y) dy$$

exists a.e. , or, equivalently,

$$\lim_{\epsilon \to 0} \int_{R^n \times A} (K_u)_\epsilon(x,y) f(y) dy d\mu$$

exists a.e. Since

$$\int_A (\sup_{\epsilon > 0} | \int_{R^n} K_\epsilon(x,y) f(y) dy |) \, d\mu$$

$$= \int_A (T_u)_* f(x) d\mu \in L^2(R^n, dx) \ ,$$

the function of x defined by the last integral is finite for almost every x , and, when it is finite, we may apply the dominated convergence theorem to obtain

$$\lim_{\epsilon \to 0} \int_{R^n \times A} (K_u)_\epsilon(x,y) f(y) dy d\mu$$

$$= \int_A [\lim_{\epsilon \to 0} \int_{R^n} K_u(x,y) f(y) dy] d\mu \ .$$

Later, we shall see two applications of this proposition; one where (A,μ) is N with the counting measure, and the other where A is a curve in the plane equipped with its arc length measure.

The following shows how the boundedness of T_* can be used even when we do not assume that the T_ϵ's approximate T :

Proposition: [CRW] Let T be a CZO and b be a function in $BMO(R^n, dx)$. Then the commutator of T and the operator of multiplication by b, $[b,T]$, is bounded on all $L^p(\omega dx)$, for $1<p<\infty$ and $\omega \in A_p$.

Note that $[b,T]$ depends only on the kernel of T, since if T_1 and T_2 have the same kernel, then they differ by a multiplication operator, and multiplication operators commute.

Let us now prove the proposition. Since the T_ϵ's are uniformly bounded, there is a sequence $\{\epsilon_k\}$ tending to zero such that T_{ϵ_k} converges weakly to some operator T_0, which must also be a CZO with the same kernel, K, as T. Hence, it suffices to show that $[b,T_0]$ has the desired boundedness properties, which can be seen to be a consequence of the following. For each $f,g \in C_c^\infty(R^n)$ such that

$$\|g\|_{L^{p'}(\omega^{-\frac{1}{p-1}}dx)} \leq 1, \quad \|f\|_{L^p(\omega dx)} \leq 1,$$

and for all $\epsilon > 0$, we have

$$\left|\int_{|x-y|>\epsilon} K(x,y)(b(x)-b(y))f(y)g(x)dydx\right| \leq C,$$

for some constant C, independent of f,g and ϵ.

To prove this, recall that $\omega \in A_p$ implies that $\omega e^a \in A_p$ for $\|a\|_*$ small enough. (This was an application of the John-Nirenberg inequality in Section I of Chapter 4; notice that its proof shows that the A_p constant of ωe^a is bounded for $\|a\|_*$ small enough.) Hence, T_ϵ is (uniformly) bounded on $L^p(\omega e^a dx)$, so that for all $h \in L^p(\omega e^a dx)$,

$$\int_{R^n} |T_\epsilon h|^p \omega(x) e^a dx \leq C \int_{R^n} |h|^p \omega(x) e^a dx .$$

Equivalently, for all $h \in L^p(\omega dx)$,

$$\int_{R^n} |T_\epsilon(he^{-\frac{a}{p}})|^p \omega(x) e^{a(x)} dx \leq C \int_{R^n} |h|^p \omega(x) dx .$$

Hence, for f and g as above, we obtain

$$\left| \int_{R^n} g(x) e^{\frac{a}{p}} T_\epsilon (f e^{-\frac{a}{p}}) dx \right| \leqq C ,$$

or

$$\left| \iint_{|x-y|>\epsilon} g(x) e^{\frac{a(x)}{p}} K(x,y) e^{-\frac{a(y)}{p}} f(y) dy dx \right| \leqq C .$$

This remains true if we replace a by za, where z describes a circle in the complex plane which is centered at the origin and has a small, but fixed, positive radius (so that $(\mathrm{Re}z)a$ has small BMO norm).

This gives us a double integral that is a holomorphic function of z (z near 0). The first order term in its power series is the expression that we wish to estimate, and the desired inequality now follows from Cauchy's formula.

The boundedness of T_* can also be used to get rid of a priori assumptions. A good example of this is given by the Cauchy kernel on a Lipschitz curve. The theorem, which we shall prove in a later chapter, is the following.

<u>Theorem</u>: <u>Let</u> $\varphi : R \to R$ <u>satisfy</u> $\|\varphi'\|_\infty < \infty$. <u>Then the kernel</u> $K_\varphi(x,y) = \frac{1}{(x-y)+i[\varphi(x)-\varphi(y)]}$ <u>defines a</u> CZSI T_φ <u>such that</u> $\|T_\varphi\|_{CZ} \leqq C(\|\varphi'\|_\infty)$.

The a priori assumption that we shall make is that $\varphi \in C_c^\infty(R)$. We can get rid of as follows:

Observe that if $f(y)$ is of the form $g(y)(1+i\varphi'(y))$ with $g \in C_c^\infty(R)$, then the existence of

$$\lim_{\epsilon \to 0} \int_{|x-y|>\epsilon} K_\varphi(x,y) f(y) dy$$

whenever φ' exists, is obvious, as one shows by integrating by parts (to take care of the singularity). Because the collection of all such f is dense in any L^p, it is enough to show that $(T_\varphi)_*$ is bounded. (Recall that such a maximal result of this type together with the a.e. convergence on a dense class is enough to insure the a.e. convergence for all $f \in L^p$.) Equivalently, it is enough to show that $(T_\varphi)_\epsilon$ is uniformly bounded. Indeed, if this is the case, then there is a sequence $\{\epsilon_K\}$ tending to zero such that $(T_\varphi)_{\epsilon_K}$ converges weakly to an operator T_φ^o. Then T_φ and T_φ^o will have the same kernel, so that $(T_\varphi)_* = (T_\varphi^o)_*$. This latter operator will be controlled by $\|T_\varphi^o\|_{2,2} \leqq \sup_{\epsilon>0} \|(T_\varphi)_\epsilon\|_{2,2}$ via Cotlar's inequality.

Now let $f, g \in C_c^\infty(R)$. Thus, we wish to prove that

$$\left| \iint_{|x-y|>\epsilon} K_\varphi(x,y)f(x)g(y)dxdy \right| \leq C(\|\varphi'\|_\infty)\|f\|_2\|g\|_2 \ .$$

To do this, we choose a sequence of functions $\varphi_j \in C_c^\infty(R^n)$ which approximates φ uniformly on $\operatorname{supp} f \cup \operatorname{supp} g$, and which satisfies $\|\varphi_j'\|_\infty \leq C\|\varphi'\|_\infty$. (To construct such a sequence is elementary and not difficult.) Hence the boundedness of the Cauchy kernel in the "regular case" yields

$$\left| \iint_{|x-y|>\epsilon} K_{\varphi_j}(x,y)f(x)g(y)dxdy \right| \leq C(\|\varphi_j'\|_\infty)\|f\|_2\|g\|_2 \ ,$$

and now the passage to the limit is immediate.

Our last application of the boundedness of T_* deals with the characterization of those convolution operators which are CZO's . (See the end of Section I.)

Assume that T is a bounded singular convolution operator with kernel K . We shall let K stand for both the kernel $K(x,y)$ of the CZO T , as well as the convolution kernel $K(x)$, so that $K(x,y) = K(x-y)$. Thus, we want to prove that

$$\sup_{0<\epsilon<\eta<\infty} \left| \int_{\epsilon<|x|<\eta} K(x)dx \right| < +\infty \ .$$

In view of the first standard estimate on K , it is enough to consider only $\epsilon < \frac{\eta}{4}$.

Because $\|T_\epsilon\|_{2,2}$ is uniformly bounded, Hölder's inequality implies that for any cube Q ,

$$\int_Q |T_\epsilon \chi_Q| dx \leq C|Q| \ ,$$

uniformly in ϵ .

Let $Q = \{x \in R^n : |x| \leq \eta\}$, so that

$$\int_{\epsilon<|x|<\eta} K(x)dx = T_\epsilon \chi_Q(0) \ .$$

For $x_0 \in \frac{1}{2}Q$,

$$\left| \int_Q K_\epsilon(x_0-y)\,dy - \int_Q K_\epsilon(-y)dy \right| \leq \int_{(Q-x_0)\Delta Q} |K_\epsilon(-y)|\,dy \ ,$$

where $A \Delta B = (A\setminus B) \cup (B\setminus A)$. Using the first standard estimate on K , one can rewrite this inequality as follows:

$$|T_\epsilon \chi_Q(x_0) - \int_{\epsilon < |x| < \eta} K(x)dx| \leqq C$$

Since

$$\frac{1}{|Q|} \int_Q |T_\epsilon \chi_Q| dx \leqq C ,$$

we obtain

$$|\int_{\epsilon < |x| < \eta} K(x)dx| \leqq C ,$$

where C is independent of ϵ and η , as desired.

CHAPTER 5

Calderón-Zygmund Operators and Pseudo-Differential Operators.

Usually, the main difficulty in showing that an operator is a Calderón-Zygmund operator lies in establishing its L^2 boundedness. We shall now consider a different type of problem. We shall study a certain family of operators, the pseudo-differential operators (ψdO's), and we wish to determine which of these are CZO's. This will lead us to a new class of L^p-bounded, $1<p<\infty$, ψdO's.

Pseudo-differential operators were introduced in the sixties by Calderón-Vaillancourt, Kohn-Nirenberg, Hörmander, and others. See [F] for bibliographical references and some applications. These operators generalize the usual differential operators, and are formally defined by a formula of the type

$$Tf(x) = \int_{R^n} e^{2\pi i\langle x,\xi\rangle}\sigma(x,\xi)\hat{f}(\xi)d\xi .$$

Note that when $\sigma(x,\xi)$ is a polynomial in ξ, with coefficients varying with x, then T is a differential operator. Also observe that we obtain a convolution operator when $\sigma(x,\xi)$ is independent of x. The function $\sigma: R^n \times R^n \to C$ is called the symbol of the operator T, and T is often written as $\sigma(x,D)$. Ordinarily, the conditions imposed on the symbol will be sufficient to insure that the above integral converges, at least when $f \in C_c^\infty(R^n)$ and $0 \notin \operatorname{supp} \hat{f}$.

In our process of identifying ψdO's which are CZO's, we shall make the a priori assumption that $\sigma \in C_c^\infty(R^n \times R^n)$. In this case, it is easy to exhibit the kernel corresponding to the operator $\sigma(x,D)$. Indeed,

$$Tf(x) = \int_{R^n}\int_{R^n} e^{2\pi i\langle x,\xi\rangle}\sigma(x,\xi)e^{-2\pi i\langle y,\xi\rangle}\hat{f}(\xi)dyd\xi$$

$$= \iint_{R^n \times R^n} e^{2\pi i\langle x-y,\xi\rangle}\sigma(x,\xi)f(y)dyd\xi .$$

Thus, writing $\sigma(x,\xi)=\sigma_x(\xi)$, we obtain $K(x,y)=\hat{\sigma}_x(y-x)$, and the standard estimates on K read as follows:

(i) $\quad \| |u|^n\hat{\sigma}_x(u)\|_\infty \leqq C$.

(ii) $\quad \| |u|^{n+1}\frac{\partial}{\partial u_j}(\hat{\sigma}_x(u))\|_\infty \leqq C, \quad j=1,2,\dots,n$.

(iii) $\| \, |u|^{n+1}\left(\frac{\partial}{\partial x_j} (\hat{\sigma}_x(u)) - \frac{\partial}{\partial u_j} (\hat{\sigma}_x(u))\right)\|_\infty \leqq C$, $j = 1,2,\ldots,n$.

Let $\alpha = (\alpha_1,\ldots,\alpha_n)$ denote an element of N^n , and let $|\alpha| = \alpha_1+\ldots+\alpha_n$. Then classical Fourier transform identies show that these three conditions are equivalent to the following three conditions:

(i)′ For $|\alpha| = n$, $\| \widehat{\frac{\partial^\alpha}{\partial \xi^\alpha} \sigma_x} \|_\infty \leqq C$.

(ii)′ For $|\alpha| = n+1$, and for $j = 1,2,\ldots,n$, $\| \widehat{\frac{\partial^\alpha}{\partial \xi^\alpha} \xi_j \sigma_x} \|_\infty \leqq C$.

(iii)′ For $|\alpha| = n+1$ and $j = 1,2,\ldots,n$, $\| \widehat{\frac{\partial^\alpha}{\partial \xi^\alpha} \frac{\partial}{\partial x_j} \sigma_x} \|_\infty \leqq C$.

Of course, we cannot hope to completely characterize the symbols $\sigma(x,D)$ for which $\sigma(x,D)$ is associated to a standard kernel, since that would be tantamount to characterizing those functions which have a bounded Fourier transform. Since we are dealing with C_c^∞ functions, we would like to know which parameters determine $\|\hat{h}\|_\infty$, for $h \in C_c^\infty$ (here h denotes any of the function in (i)′, (ii)′, (iii)′). Thus, the following result will be useful: see [S]:

<u>Lemma</u>: <u>If</u> $h \in C_c^\infty(R^n)$ <u>satisfies</u> $|h(y)| \leqq \frac{C(h)}{|y|^n}$ <u>and</u> $|\nabla h| \leqq \frac{C(h)}{|y|^{n+1}}$ <u>for all</u> $y \in R^n$, <u>and</u>

$$\sup_{0<R_1<R_2} \left| \int_{R_1<|y|<R_2} h(y)dy \right| \leqq C(h) ,$$

<u>then</u> $\|\hat{h}\|_\infty \leqq c_n C(h)$

This lemma follows from the results in the preceding chapter. For if $K(x,y) = h(x-y)$, then the above conditions on h imply that K is the kernel of an operator T that is a CZO . Hence $Tg = \hat{h}\hat{g}$, and since T is bounded on L^2 , $\|\hat{h}\|_\infty \leqq \|T\|_{CZ}$.

If we apply this lemma to each of the functions appearing in (i)′, (ii)′ and (iii)′ above, then we obtain the following result.

<u>Proposition</u>: <u>If</u> $\sigma \in C_c^\infty(R^n \times R^n)$ <u>satisfies</u>

(1) $$\left| \frac{\partial^\alpha}{\partial \xi^\alpha} \frac{\partial^\beta}{\partial x^\beta} \sigma(x,\xi) \right| \leqq C_{\alpha,\beta} |\xi|^{|\beta| - |\alpha|}$$

<u>for</u> $|\alpha| = 0,1,\ldots,n+2$, $|\beta| = 0,1$, <u>and</u> $|\beta| \leqq |\alpha|$, <u>then</u> $\sigma(x,D)$ <u>is associated to a standard kernel</u> K , <u>where</u> $C(K)$ <u>depends only on the constants</u> $C_{\alpha\beta}$.

To prove this, we must show that the functions satisfying (i)′, (ii)′, and (iii)′ satisfy the hypotheses of the lemma. Notice first that (i)′ and (ii)′ together are equivalent to (i)′ and (ii)′′ , where (ii)′′ is the following condition. for $|\alpha| = n+1$ and $j = 1,\ldots,n$,

$$\|\xi_j \frac{\partial^\alpha}{\partial \xi^\alpha} \sigma_x\|_\infty \leqq C .$$

Using the fact that $\frac{1}{|\xi|} \approx \inf_{1 \leqq j \leqq n} \frac{1}{|\xi_j|}$, it follows immediately that a function which satisfies (i)′, (ii)′, and (iii)′ also satisfies the first two hypotheses of the lemma. (There we consider h as a function of ξ , x being held fixed.)

To check the third hypothesis, once again we suppose that the norm on R^n which we are working with is such that its unit ball is a cube. (Easy estimates show that the difference between this norm and, say, the Euclidian norm yields appropriately controllable error terms.) For $|\alpha| = n$, we estimate

$$\int_{R_1 < |\xi| < R_2} \frac{\partial^\alpha}{\partial \xi^\alpha} \sigma_x \, d\xi$$

by integrating with respect to one of the variables of the n-tuple α , say x_i . Then it is easy to show that the uniform boundedness of the integral follows from

$$\left| \frac{\partial^{\alpha'}}{\partial \xi^{\alpha'}} \sigma_x \right| \leqq C_{\alpha'} |\xi|^{1-n} , \quad |\alpha'| = n-1 .$$

The symbols considered in the above proposition are (aside from the a priori assumption) more general than the classes $S^0_{1,\delta}$, $0 \leqq \delta < 1$, introduced by Hörmander in [H] . Unfortunately, these symbols (i.e., the ones which satisfy condition (1) in the above proposition) do not always lead to bounded operators on L^2 , as is shown by the following one-dimensional counterexample, communicated to us by Yves Meyer.

<u>Counterexample</u>: <u>Let</u> $\varphi \in C_c^\infty(R)$ <u>be zero outside of</u> $[\frac{11}{21}, \frac{19}{21}]$ <u>and equal to</u> 1 <u>on</u> $[\frac{4}{7}, \frac{6}{7}]$. <u>Then the symbol</u>

$$\sigma(x,\xi) = \sum_{n=0}^{+\infty} e^{-2\pi i \frac{5}{7} 2^{-n} x} \varphi(2^n \xi)$$

satisfies (1) above, and, moreover, the formula

$$Tf(x) = \int_{\mathbb{R}} \sigma(x,\xi)e^{2i\pi x\xi}\hat{f}(\xi)d\xi$$

is meaningful, at least for $\hat{f}$ vanishing near 0, but T cannot be extended to a bounded operator on L^2.

Proof: To see that σ satisfies (1), notice that the functions

$$\sigma_n = e^{-2\pi i \frac{5}{7}2^{-n}x}\varphi(2^n\xi)$$

have disjoint supports, since $\operatorname{supp}\varphi(2^n\xi)$ is contained in $[2^{-n-1}, 2^{-n})$. Hence it is enough to show that σ_n satisfies (1) uniformly in n, which is an immediate consequence of the fact that $(x,\xi) \in \operatorname{supp}\sigma_n$ implies that $2^n \leqq \frac{2}{|\xi|}$.

We now show that there exists no constant K such that for each $f \in L^2(\mathbb{R})$, $0 \notin \operatorname{supp}\hat{f}$, $\|Tf\|_2 \leqq K\|f\|_2$.

Let $f \in L^2(\mathbb{R})$ be such that $\hat{f}$ is bounded with a compact support not containing 0. Then,

$$Tf(x) = \int_{\mathbb{R}} \sum_{n=0}^{\infty} e^{-2\pi i \frac{5}{7}2^{-n}x}\varphi(2^n\xi)\hat{f}(\xi)e^{2\pi i x\xi}d\xi ,$$

where the sum has only finitely many nonvanishing terms, since $0 \notin \operatorname{supp}\hat{f}$. Therefore,

$$\widehat{Tf} = \sum_{n=0}^{\infty} \varphi(2^n\xi + \tfrac{5}{7})\hat{f}(\xi + \tfrac{5}{7}2^{-n}) .$$

For $\xi \in [\frac{2}{3}(2^{-n}\frac{1}{7}), (2^{-n}\frac{1}{7})]$, it is easy to check that $\varphi(2^p\xi + \frac{5}{7}) = 1$ if $p \leqq n$ and vanishes if $p > n$. Hence, for such ξ,

$$\widehat{Tf}(\xi) = \sum_{p=0}^{n} \hat{f}(\xi + \tfrac{5}{7}2^{-p}) .$$

We now define $\hat{f}$ as follows. Let $a_{n,q}$ be a positive sequence on $\mathbb{N} \times \mathbb{N}$, vanishing when n or q is large, and when $n < q$. For $n \geqq q$ set $\hat{f}(\xi) = a_{nq}$ if $\frac{2}{3}\cdot\frac{2^{-n}}{7} \leqq |\xi - \frac{5}{7}2^{-q}| \leqq \frac{2^{-n}}{7}$ and $\hat{f}(\xi) = 0$ otherwise. Since we need only consider $n \geqq q$, it is easy to see that the sets we just defined are disjoint. The estimates on $\|f\|_2$ and $\|Tf\|_2$ are now straight forward, and we obtain

$$\|f\|_2^2 = \sum_{q=0}^{\infty} \sum_{n=q}^{\infty} \frac{2}{21} \cdot 2^{-n} (a_{nq})^2$$

and

$$\|Tf\|_2^2 \geqq \sum_{n=0}^{\infty} \left(\frac{1}{21} \cdot 2^{-n}\right) \left(\sum_{q=0}^{n} a_{nq}\right)^2 .$$

If we let $a_{nq} = \dfrac{2^{n/2}}{(n+1)^{5/4}}$ for $n \geqq q$, $a_{nq} = 0$ if $n < q$, then

$$\sum_{n=0}^{\infty} 2^{-n} \left(\sum_{q=0}^{n} a_{nq}^2\right) < \infty$$

and

$$\sum_{n=0}^{\infty} 2^{-n} \left(\sum_{q=0}^{n} \;_{nq}\right)^2 = \infty .$$

Hence, there cannot exist a constant K such that $\|Tf\|_2 \leqq K\|f\|_2$ for all $f \in L^2(\mathbb{R})$.

Note that the symbol

$$\sigma'(x,\xi) = \sum_{n=0}^{\infty} e^{-2\pi i \frac{5}{7} 2^n x} \varphi(2^{-n}\xi)$$

could also have been chosen as a counterexample.

In view of this counterexample, we must strengthen (1) above in order to have $\sigma(x,D)$ bounded on L^2 . The condition (1)′,

$$(1)' \qquad \left| \frac{\partial^\alpha}{\partial\xi^\alpha} \frac{\partial^\beta}{\partial x^\beta} \sigma(x,\xi) \right| \leqq C_{\alpha,\beta} (1+|\xi|)^{|\beta| - |\alpha|} ,$$

is not enough either, in view of the second counterexample, σ' . In fact, what we need is

$$(1)'' \qquad \left| \frac{\partial^\alpha}{\partial\xi^\alpha} \frac{\partial^\beta}{\partial x^\beta} \sigma(x,\xi) \right| \leqq C_{\alpha,\beta} (1+|\xi|)^{\delta|\beta| - |\alpha|} ,$$

for some δ , $0 \leqq \delta < 1$.

<u>Theorem</u>: <u>Suppose that</u> σ <u>satisfies</u> (1)″ <u>for</u> $|\alpha| \leqq n+1$, $|\beta| \leqq 1$, <u>and</u> $|\alpha| \geqq |\beta|$. <u>Then</u> $\sigma(x,D)$ <u>is bounded on</u> $L^2(\mathbb{R}^n)$.

<u>Proof</u>: We shall follow closely a proof of Coifman and Meyer in [CM2].

Let us first consider a special type of symbol; the <u>reduced symbols</u>, which are the ones having the form

$$\sigma(x,\xi) = \sum_{j=0}^{\infty} m_j(x)\varphi(2^{-j}\xi) \quad ,$$

where $m_j : R^n \to R$ satisfies

$$\|m_j\|_\infty \leq C_1 \ , \ \|\nabla_x m_j\|_\infty \leq C_1 2^{j\delta}, \ 0 \leq \delta < 1 \ ,$$

and $\varphi \in C^{n+1}(R^n)$ is supported in $\{\xi : \frac{1}{3} \leq |\xi| \leq 3\}$. Notice that

$$\operatorname{supp} \varphi(2^{-j}\cdot) \cap \operatorname{supp} \varphi(2^{-k}\cdot) = \emptyset$$

if $|j-k| \geq 4$.

Lemma 1: A reduced symbol is bounded on L^2, with operator norm dominated by a constant depending only on C_1, δ, and $\|\varphi\|_\infty$.

To prove the lemma, let $f \in C_c^\infty(R^n)$, and for $j \in N$, define $\hat{f}_j$ by $\hat{f}_j = f\varphi(2^{-j}\cdot)$. Now,

$$\sigma(x,D)(f)(x) = \int_{R^n} \sum_{j=0}^{\infty} m_j(x)\varphi(2^{-j}\xi)e^{2\pi i\langle x,\xi\rangle}\hat{f}(\xi)d\xi$$

$$= \sum_{j=0}^{\infty} m_j(x)f_j(x) \ ,$$

where the above "double integral" (over $R^n \times N$) is absolutely convergent. From the "overlap" property of the $\varphi(2^{-j}\cdot)$ together with Plancherel's theorem we conclude that

$$\sum_{j=0}^{\infty} \|f_j\|_2^2 \leq 4\|\varphi\|_\infty^2 \|f\|_2^2 \ .$$

Hence, it is enough to prove

(2) $$\|\sum_{j=0}^{\infty} m_j(x)f_j(x)\|_2 \leq C_\delta C_1 (\Sigma \|f_j\|_2^2)^{\frac{1}{2}} \ .$$

This will be a consequence of the following.

Lemma 2: There exists a positive constant C_2, depending only on C_1 and on the dimension n, such that $m_j = g_j + b_j$, where $\|g_j\|_\infty \leq C_2, \|b_j\| \leq C_2 2^{(\delta-1)j}$, and

$$\operatorname{supp} \widehat{f_j g_j} \subseteq \{ \tfrac{1}{6} 2^j \leqq |\xi| \leqq 2^j (3 + \tfrac{1}{6}) \} .$$

To show this, we let $K \in C_c^\infty(R^n)$ be such that $\hat{K}(0) = 1$ and $\hat{K} = 0$ on $|\xi| > 1$, and we set $K_j(x) = (\frac{2^j}{6})^n K(\frac{2^j x}{6})$ for $j \in N$. Then the decomposition $g_j = K_j * m_j$ and $b_j = m_j - g_j$ satisfies the conclusions of the lemma.

Clearly,

$$\|g_j\|_\infty \leqq \|K_j\|_1 \|m_j\|_\infty \leqq \|K_j\|_1 C_1 .$$

To estimate $\|b_j\|_\infty$ we use the fact that $1 = \hat{K}_j(0) = \int K_j dx$, so that

$$|b_j(x)| = |\int_{R^n} [m_j(x) - m_j(x - t)] K_j(t) dt|$$

$$\leqq \int_{R^n} C_1 |t| 2^{j\delta} (\frac{2^j}{6})^n |K(\frac{t 2^j}{6})| dt$$

$$\leqq C_1 \cdot 6 \cdot 2^{j(\delta - 1)} \int_{R^n} |t| \, |K(t)| dt .$$

The last assertion, concerning $\operatorname{supp} \widehat{f_j g_j}$, is immediate when $m_j \in L^2$, since then $g_j \in L^2$ and $\operatorname{supp} g_j \subseteq \operatorname{supp} K_j$, and, thus,

$$\operatorname{supp} \widehat{f_j g_j} \subseteq \operatorname{supp} \hat{f}_j + \operatorname{supp} \hat{g}_j$$

$$\subseteq \{ \frac{2^j}{3} \leqq |\xi| \leqq 3 \cdot 2^j \} + \{ |\xi| \leqq \frac{2^j}{6} \}$$

$$\subseteq \{ \frac{2^j}{6} \leqq |\xi| \leqq (3 + \tfrac{1}{6}) 2^j \} .$$

We obtain the general case, when $m_j \notin L^2$, by approximation m_j by $m_j \chi_{\{|\xi| < p\}} = m_j \chi_p$, and letting $p \to \infty$. Indeed, the dominated convergence theorem implies that

$$\lim_{p \to \infty} \int_{R^n} [f_j (K_j * (m_j \chi_p)) - f_j (K_j * m_j)]^2 dx = 0 ,$$

and then Plancherel's theorem implies that

$$\operatorname{supp}(f_j(K_j * m_j)) \subseteq \limsup_{p\to\infty}\,[\operatorname{supp}(f_j(K_j*(m_j\chi_p)))]$$

$$\subseteq \{\tfrac{2^j}{6} \leq |\xi| \leq (3+\tfrac{1}{6})\,2^j\} .$$

To finish the proof of Lemma 1, we shall estimate $\|\sum_{j=0}^{\infty} b_j(x)f_j(x)\|_2$ and $\|\sum_{j=0}^{\infty} g_j(x)f_j(x)\|_2$.

$$\|\sum_{j=0}^{\infty} b_j f_j\|_2 \leq \sum_{j=0}^{\infty} \|b_j f_j\|_2$$

$$\leq \sum_{j=0}^{\infty} \|b_j\|_\infty \|f_j\|_2$$

$$\leq (\sum_{j=0}^{\infty} \|b_j\|_\infty^2)^{\frac{1}{2}} (\sum_{j=0}^{\infty} \|f_j\|_2^2)^{\frac{1}{2}}$$

$$\leq C_\delta \cdot C_1 (\sum_{j=0}^{\infty} \|f_j\|_2^2)^{\frac{1}{2}} .$$

Notice the importance of the restriction $\delta < 1$, used to control $\sum_{j=0}^{\infty} \|b_j\|^2$. We estimate $\|\Sigma g_j f_j\|_2^2$ using the fact that $|j-k| \geq 5$ implies

$$\operatorname{supp} \widehat{f_j g_j} \cap \operatorname{supp} \widehat{f_k g_k} = \emptyset$$

Hence, by Plancherel's theorem,

$$\|\sum_{j=0}^{\infty} f_j g_j\|_2 \leq \sum_{i=0}^{4} \|\sum_{k=0}^{\infty} (fg)_{i+5k}\|_2$$

$$\leq 5 (\sum_{j=0}^{+\infty} \|f_j g_j\|_2^2)^{\frac{1}{2}}$$

$$\leq 5\|K\|_1 C_2 [\sum_{j=0}^{\infty} \|f_j\|_2^2]^{\frac{1}{2}} .$$

This completes the proof of Lemma 1.

Our next task is to write our general symbol as a sum of reduced symbols, plus another symbol vanishing on $|\xi| > 1$.

Let $\lambda \in C_c^\infty(\mathbb{R}^n)$ be supported in $\{\frac{1}{3}+\epsilon \leqq |\xi| \leqq 1-\epsilon\}$, for some $\epsilon \in (0,\frac{1}{12})$, and such that

$$\sum_{j=-\infty}^{\infty} \lambda(2^{-j}\xi) = 1 \quad \text{if} \quad \xi \neq 0 \ .$$

Then we write, when $\sigma(x,\xi)$ is a general symbol satisfying $\sigma(x,0)=0$,

$$\begin{aligned}\sigma(x,\xi) &= \Big(\sum_{j=-\infty}^{0} \lambda(2^{-j}\xi)\Big)\sigma(x,\xi) \\ &\quad + \sum_{j=1}^{\infty} \lambda(2^{-j}\xi)\sigma(x,\xi) \\ &= \tau(x,\xi) + \sum_{j=1}^{\infty} a_j(x,\xi) \ .\end{aligned}$$

The symbol $\tau(x,\xi)$ vanishes for $|\xi| \geqq 1$, and there exists a constant C depending on λ and on the constants $C_{\alpha\beta}$ (in condition (1)″) such that

$$\Big\| \frac{\partial^\alpha}{\partial\xi^\alpha} \tau(x,\xi)\Big\|_\infty \leqq C$$

if $|\xi| \leqq 1$ and $|\alpha| \leqq n+1$.

Because a bounded function with compact support has a bounded Fourier transform, it follows from classical Fourier transform formulas that $\hat{\tau}_x(u)$ is bounded both by C and by $\frac{C}{|u|^{n+1}}$. Thus, $\tau(x,D)$ is dominated by the operator of convolution with $\frac{1}{1+|u|^{n+1}}$, and is therefore bounded on $L^2(\mathbb{R}^n)$.

To control the a_j's we shall need the following:

<u>Lemma 3</u>: <u>Let</u> $\psi \in C_c^\infty(\mathbb{R}^n)$ <u>be such that</u> $\psi\lambda = \lambda$ <u>and</u> $\operatorname{supp}\psi \subseteq \{\frac{1}{3} \leqq |\xi| \leqq 1\}$. <u>Then there exist functions</u> $a_{jk}, j,k \in N$, <u>such that</u>

(i) $\|a_{jk}\|_\infty \leqq C$;

(ii) $\|\nabla a_{j,k}\|_\infty \leqq C2^{j\delta}$;

(iii) $a_j(x,\xi) = \sum_{k\in Z^n} (1+|k|^2)^{-\frac{n+1}{2}} a_{jk}(x) e^{i\langle k,\xi\rangle 2^{-j}} \psi(\xi 2^{-j})$.

Before proving Lemma 3, let us show how it permits us to write $\sigma - \tau$ as a convergent series of reduced symbols. We choose ψ, as above. We then have:

$$\sum_{j=1}^{\infty} a_j(x,\xi) = \sum_{j=1}^{\infty} \sum_{k\in Z^n} (1+|k|^2)^{-\frac{n+1}{2}} a_{jk}(x)e^{i\langle k,\xi\rangle 2^{-j}}\psi(\xi 2^{-j}) ,$$

where the double series converges absolutely, since the functions $\psi(2^{-j}\cdot)$ have finitely overlapping supports. Define σ_k by

$$\sigma_k(x,\xi) = \sum_{j=1}^{\infty} a_{jk}(x)e^{i\langle k,\xi\rangle 2^{-j}}\psi(\xi 2^{-j}) ,$$

which is a reduced symbol.

Because the operator norm estimate in Lemma 1 depends only on C_1, δ, and $\|\varphi\|_\infty$, and not on the derivatives of φ, we see that the operator norm of $\sigma_k(x,D)$ on L^2 is uniformly bounded.

Now,

$$\sigma(x,\xi) = \tau(x,\xi) + \sum_{k\in Z^n} \frac{1}{(1+|k|^2)^{\frac{n+1}{2}}} \sigma_k(x,\xi) ,$$

and for $f \in C_c^\infty(R^n)$, we can express $\sigma(x,D)(f)(x)$ by

$$\int_{R^n} \sigma(x,\xi)e^{2\pi i\langle x,\xi\rangle}\hat{f}(\xi)d\xi .$$

For such f it is clear that

$$\sigma(x,D)f = \tau(x,D)f + \sum_{k\in Z^n} \frac{1}{(1+|k|^2)^{\frac{n+1}{2}}} \sigma_k(x,D)f .$$

Since $C_c^\infty(R^n)$ is dense in $L^2(R^n)$, we obtain

$$\|\sigma(x,D)\|_{2,2} \leqq \|\tau(x,D)\|_{2,2} + \sum_{k\in Z^n} \frac{\|\sigma_k(x,D)\|_{2,2}}{(1+|k|^2)^{\frac{n+1}{2}}}$$

$$\leqq C(\delta, c_{\alpha,\beta}) .$$

<u>Proof of Lemma 3</u>: Define a function A_j on $R^n \times R^n$ by

$$A_j(x,\xi) = \sum_{k\in Z^n} a_j(x,2^j(\xi - 2k\pi)) .$$

Because $a_j(x,2^j\xi)$ vanishes for $|\xi|>1$, this is meaningful and defines a function which is 2π-periodic in the ξ co-ordinates. Also, since $\psi\lambda = \lambda$,

$$a_j(x,\xi) = \psi(\xi 2^{-j})a_j(x,\xi) = \psi(\xi 2^{-j})A_j(x,\xi 2^{-j}) .$$

We can write A_j as the sum of its Fourier series:

$$A_j(x,\xi) = \sum_{k\in Z^n} C_{j,k}(x)e^{i\langle k,\xi\rangle} ,$$

where

$$C_{j,k}(x) = (2\pi)^{-n} \int_{R^n} e^{-i\langle\xi,k\rangle} a_j(x,2^j\xi)d\xi .$$

Let $a_{jk}(x) = (1+|k|^2)^{\frac{n+1}{2}} C_{jk}(x)$. Then (iii) is immediate, and we must show that

(i)′ $\quad \|C_{jk}\|_\infty \leqq \dfrac{C}{(1+|k|^2)^{\frac{n+1}{2}}}$

(ii)′ $\quad \|\nabla C_{jk}\|_\infty \leqq \dfrac{C}{(1+|k|^2)^{\frac{n+1}{2}}} 2^{j\delta}$.

To prove (i)′, we use the formula defining $C_{jk}(x)$ and integrate by parts $n+1$ times in one of the co-ordinates, say ξ_s, $1\leqq s\leqq n$. Thus,

$$C_{jk}(x) = (2\pi)^{-n} \int_{R^n} \frac{(-1)^{n+1}}{(-ik_s)^{n+1}} e^{-i\langle\xi,k\rangle} \frac{\partial^{n+1}}{\partial\xi_s^{n+1}} \sigma_j(x,2^j\xi) \quad (2^j)^{n+1}d\xi .$$

Since the a_j's satisfy (i)″ uniformly in j,

$$\|C_{j,k}\|_\infty \leqq \frac{C}{|k_s|^{n+1}} \int_{\mathrm{supp}(a_j(x,2^j\,\cdot\,))} \frac{1}{|\xi|^{n+1}} d\xi \leqq \frac{C}{|k_s|^{n+1}} .$$

Since this is true for all s, and since $\|C_{jk}\|_\infty \leqq C$, it follows that

$$\|c_{jk}\|_\infty \leq \frac{C}{(1+|k|^2)^{\frac{n+1}{2}}},$$

which is (i)′. One proves (ii)′ similarly.

This completes the proof of Lemma 3, and hence of the theorem. We shall combine the theorem and the earlier proposition to obtain the following.

Theorem: Suppose that $\sigma : R^n \times R^n \to R$ is $C^{n+3}(R^n \times R^n)$, and suppose that there is a $\delta > 0$ such that for $\alpha, \beta \in N^n$, $|\alpha| \leq n+2$, $|\beta| \leq \inf(1, |\alpha|)$, there exists $C_{\alpha\beta}$ such that

$$\left| \frac{\partial^\alpha}{\partial \xi^\alpha} \frac{\partial^\beta}{\partial x^\beta} \sigma(x,\xi) \right| \leq C_{\alpha\beta}(1+|\xi|)^{\delta|\beta| - |\alpha|}.$$

Then $\sigma(x,D)$ is a CZO.

To prove this theorem, we must rid ourselves of the a priori assumptions.

Recall that in the proposition we assumed that $\sigma \in C_c^\infty(R^n \times R^n)$. In fact, a quick review of the proof shows that all we need to assume is that $\sigma \in C^{n+3}(R^n \times R^n)$ and that its support in ξ is contained in some fixed compact set which is independent of x. Let us now indicate how one can get rid of this compactness assumption.

Let $\varphi \in C_c^\infty(R^n)$ be equal to 1 near the origin, and suppose that σ satisfies the hypothesis of the theorem. Then $\sigma_j(x,\xi) = \varphi(\frac{\xi}{j})\sigma(x,\xi)$ has uniformly compact support in ξ, and satisfies the hypotheses of the theorem uniformly in j. Moreover, the σ_j approximate σ in the following sense

Lemma: For $\psi \in C_c^\infty(R^n)$,

$$\lim_{j\to\infty} \|(\sigma_j(x,D) - \sigma(x,D))\psi\|_2 = 0$$

Proof: For $\psi \in \mathcal{S}(R^n)$, let $\psi_j = (\varphi(\frac{\xi}{j})\hat{\psi})^\vee$. Then

$$\sigma_j(x,D)\psi = \sigma(x,D)\psi_j,$$

and, hence,

$$\lim_{j\to\infty} \|[\sigma_j(x,D) - \sigma(x,D)]\psi\|_2 \leq \|\sigma(x,D)\|_{2,2} \lim_{j\to\infty} \|\psi - \psi_j\|_2$$

Finally, the theorem is a consequence of the following general property of CZO's, see [CM2] for a proof.

Proposition: suppose that T is a bounded linear operator on L^2 and that there exists a sequence $(T_j)_{j\in N}$ of CZO's such that

(i) $\|T_j\|_{CZ}$ is uniformly bounded in j ;

and

(ii) $\lim_{j\to\infty}\|T_j\psi - T\psi\|_{2,2} = 0$ for all $\psi \in \mathcal{S}(\mathbb{R}^n)$.

Then T is a CZO .

Another application of this proposition is the following.

Corollary: Let $\sigma : \mathbb{R}^n \to \mathbb{R}$ be of class $C^{n+3}(\mathbb{R}^n)$ and satisfy

$$\left|\frac{\partial^\alpha}{\partial \xi^\alpha}\sigma(\xi)\right| \leq C_\alpha |\xi|^{-|\alpha|}$$

for all $\alpha \in N^n$, $|\alpha| \leq n+2$. Then σ defines a convolution operator which is a CZO .

To prove the corollary, observe that the result follows from the theorem if σ vanishes on $\{|\xi| \leq 1\}$, since $1 \leq \frac{1+|\xi|}{|\xi|} \leq 2$ if $|\xi| > 1$.

Note that if σ defines a CZO T , then $\sigma(r\,\cdot)$ defines a CZO T_r for $r>0$, and $\|T_r\|_{CZ} = \|T\|_{CZ}$. Indeed, the kernel of T is $K(x,y) = \sigma(y-x)$, and the kernel of T_r is $K_r(x,y) = \frac{1}{r^n}\sigma\left(\frac{y-x}{r}\right)$, so that if K satisfies the standard estimates, then so does K_r , with the same constant. Moreover, $\|T_r\|_{2,2} = \|T\|_{2,2} = \|\sigma\|_\infty$.

By the preceding remark, the case where σ vanishes near the origin now follows, by dilating, from the case where σ vanishes on $\{|\xi| \leq 1\}$. We now apply the previous proposition to the symbols $\sigma_j = \sigma(1-\varphi(\xi_j))$, where $\varphi \in C_c^\infty(\mathbb{R}^n)$ and φ equals 1 near the origin. Thus, for $\psi \in \mathcal{S}(\mathbb{R}^n)$,

$$\begin{aligned} &\lim_{j\to\infty}\|[\sigma(x,D) - \sigma_j(x,D)]\psi\|_2 \\ &\leq \|\sigma\|_\infty \lim_{j\to\infty}\|\varphi(\xi_j)\psi\|_2 = 0 , \end{aligned}$$

by Plancherel's theorem. The corollary now follows from the proposition.

All of the results and methods of this chapter are valid in a more general vector-valued context.

Let H_1, H_2 be Hilbert spaces. For a function $f \in L^1_{H_1}(\mathbb{R}^n)$, define

$$\hat{f}(\xi) = \int_{\mathbb{R}^n} e^{-2\pi i\langle x,\xi\rangle} f(\xi)d\xi .$$

Clearly $\|\hat{f}\|_\infty \leq \|f\|_1$. By using an orthonormal basis of H_1 one can easily show that $\|\hat{f}\|_2 = \|f\|_2$ for $f \in L^1_{H_1}(\mathbb{R}^n) \cap L^2_{H_1}(\mathbb{R}^n)$.

Let $\sigma(x,\xi) : \mathbb{R}^n \times \mathbb{R}^n \to \mathcal{B}(H_1,H_2)$ satisfy suitable assumptions. Then we can define an operator $\sigma(x,D)$ from $C^\infty_{c,H_1}(\mathbb{R}^n)$ into $L^1_{loc,H_2}(\mathbb{R}^n)$ by

$$\sigma(x,D)(f)(x) = \int_{\mathbb{R}^n} \sigma(x,\xi)\hat{f}(\xi)e^{2\pi i\langle x,\xi\rangle}d\xi .$$

Then the appropriate analogues of the preceding results hold.

An application of this is the following.

Let $H_1 = C$ and $H_2 = \ell^2$, and suppose that $\varphi \in C^\infty_c(\mathbb{R}^n)$ is 1 on $[\frac{1}{2},1]$ and 0 outside of $[\frac{1}{4},2]$. Then we define $\sigma(\xi) = (\varphi(2^k\xi))_{k\in\mathbb{Z}}$ (note that, since $H_1 = C$, $\mathcal{B}(H_1,H_2)$ is essentially the same as H_2). Thus, σ satisfies the hypothesis of the corollary, and if we write $f_k = [\hat{f}\varphi(2^k\cdot)]^\vee$ for $f \in L^2$, then we can conclude that $f \to (f_k)_{k\in\mathbb{Z}}$ defines a (vector-valued) CZO. In particular,

$$\| [\sum_{k=-\infty}^{+\infty} |f_k|^2]^{\frac{1}{2}} \|_p \leq C\|f\|_p .$$

This can be used in proving the dyadic decomposition theorem of Littlewood and Paley. See Chapter 6, as well as [S].

CHAPTER 6

CALDERÓN-ZYGMUND OPERATORS AND LITTLEWOOD-PALEY THEORY.

I. Introduction.

We are now going to apply the techniques developed so far to study certain types of operators which arise in Littlewood-Paley theory. We shall not try to describe, even briefly, what Littlewood-Paley theory is all about, however; a detailed exposition may be found in [CW2].

By a "Littlewood-Paley operator" we shall mean the following kind of object. Let (A,μ) be a measure space, and for each $u \in A$, let T_u be a linear operator acting on $\mathcal{S}(\mathbb{R}^n)$ such that for all $f \in \mathcal{S}(\mathbb{R}^n)$

$$f \sim \int_A (T_u f) d\mu(u) ,$$

where the relation "$\sim$" will often mean equality, but not always (see the examples below). Then we define the operator T by

$$Tf = [\int_A (T_u f)^2 d\mu(u)]^{\frac{1}{2}} ,$$

and we say that it is a Littlewood-Paley operator if it is an isometry on L^2. Clearly if two functions f and g satisfy $|T_u f| = |T_u g|$ for each $u \in A$, then $Tf = Tg$. The isometric property has an important easy consequence, applications of which we shall give later.

Proposition: If T is bounded on $L^p(\omega dx)$ for some $p \in (1,\infty)$ and $\omega \in A_p$, then T satisfies the following converse inequality: for $f \in L^{p'}(\omega^{-\frac{1}{p-1}} dx)$

$$\|f\|_{L^{p'}(\omega^{-\frac{1}{p-1}} dx)} \leq C(\omega,p) \|Tf\|_{L^{p'}(\omega^{-\frac{1}{p-1}} dx)} .$$

where $\frac{1}{p'} + \frac{1}{p} = 1$.

(If T is bounded on $L^p(\omega dx)$ for all $p \in (1,\infty)$ and $\omega \in A_p$, then Tf and f have equivalent $L^p(\omega dx)$ norms.)

Proof: (See [S].) Let $f, g \in \mathcal{S}(\mathbb{R}^n)$ be given. Then, since T is an isometry,

$$\|T(f+g)\|_2^2 - \|Tf\|_2^2 - \|Tg\|_2^2 = \|f+g\|_2^2 - \|f\|_2^2 - \|g\|_2^2 ,$$

which reduces immediately to

$$\int_{R^n}\int_A (T_u f)(T_k g)d\mu dx = \int_{R^n} fgdx .$$

If we use Schwarz's inequality on the integral with respect to μ , we obtain

$$|\int_{R^n} fgdx| \leq \int_{R^n}(Tf)(Tg)dx .$$

An application of Hölder's inequality to the right hand side gives us

$$|\int_{R^n} fgdx| \leq [\int_{R^n}|Tf|^p \omega dx]^{\frac{1}{p}}[\int_{R^n}|Tg|^{p'}\omega^{-\frac{1}{p-1}}dx]^{\frac{p-1}{p}} ,$$

Hence,

$$|\int_{R^n} fgdx| \leq C[\int_{R^n}|f|^p \omega dx]^{\frac{1}{p}}[\int_{R^n}|Tg|^{p'}\omega^{-\frac{1}{p-1}}dx]^{\frac{p-1}{p}} .$$

If we take the supremum of this over the $f \in \mathcal{S}(R^n)$ such that $\|f\|_{L^p(\omega dx)} = 1$, we obtain

$$[\int_{R^n}|g|^{p'}\omega^{-\frac{1}{p-1}}dx]^{\frac{1}{p'}} \leq C[\int_{R^n}|Tg|^{p'}\omega^{-\frac{1}{p-1}}dx]^{\frac{1}{p'}} .$$

The assumption that $g \in \mathcal{S}(R^n)$, can be easily removed since T is positive and sublinear.

II. The Littlewood-Paley G - and S - Functions.

The following is an example of a classical Littlewood-Paley operator: Let $h:R^n \to R$ satisfy

(i) $\int_{R^n} h(t)dt = 0$, and

(ii) there exists $C_h > 0$ such that for $x \in R^n$, $|h(x)| \leq \frac{C_h}{(1+|x|)^{n+1}}$ and $|\nabla h(x)| \leq \frac{C_h}{(1+|x|)^{n+2}}$. Let $(A,\mu) = (R_+, \frac{dt}{t})$ and $T_t f(x) = (h_t * f)(x)$, where, as usual, $h_t(x) = \frac{1}{t^n}h(\frac{x}{t})$. We let H denote the Hilbert space $L^2(R_+, \frac{dt}{t})$, and with T as in the preceding section, we define $\tilde{T}f(x) = g_x \in H$, where $g_x(t) = (h_t * f)(x) = T_t f(x)$. Hence $Tf(x) = \|\tilde{T}f(x)\|_H$, and T and $\tilde{T}$ must therefore have the same boundedness properties.

<u>Theorem</u>: <u>$\tilde{T}$ is a $(\mathbb{C}, H)$ CZO, and, if $|\hat{h}|$ is a radial function, then T is a Littlewood-Paley operator.</u>

Actually, if $|\hat{h}|$ is radial, then T is an isometry only up to a constant factor, which is a distinction that we shall ignore.

<u>Proof</u>: We identify again $\mathcal{B}(\mathbb{C}, H)$ with H. It is easy to guess that the kernel of $\tilde{T}$, which takes its values in this space, is given by $K(x,y)(t) = h_t(x-y)$. We shall justify the correctness of this guess.

The size assumptions on h and ∇h imply that K, defined as above, is a standard kernel. Indeed, the first standard estimate,

$$\left(\int_0^\infty |h_t(x-y)|^2 \frac{dt}{t}\right)^{\frac{1}{2}} \leq \frac{C_h}{|x-y|^n}$$

follows immediately from the first part of (ii) above, and the second standard estimate is dealt with similarly.

Let us now show that if $f \in L_c^\infty(\mathbb{R}^n)$ and if $x \notin \operatorname{supp} f$, then

$$\tilde{T}f(x) = \int_{\mathbb{R}^n} K(x,y) f(y) dy .$$

The first standard estimate on K shows that $K(x,\cdot)f(\cdot) \in L_H^1(\mathbb{R}^n, dy)$ for $x \notin \operatorname{supp} f$, and so we must prove

$$\left(\int_{\mathbb{R}^n} K(x,y) f(y) dy\right)(t) = \int_{\mathbb{R}^n} K(x,y)(t) f(y) dy$$

for almost every $t \in \mathbb{R}_+$. The problem in doing this is that the operator $\pi_t : H \to \mathbb{R}$ defined by $\pi_t(f) = f(t)$ is not a well-defined continuous functional, so that we cannot directly apply the properties of the Bochner integral. Fortunately, however, we can approximate π_t by continuous functionals.

For $0 < t_1 < t_2 < \infty$, define π_{t_1,t_2} by

$$\pi_{t_1,t_2}(g) = \frac{1}{t_2 - t_1} \int_{t_1}^{t_2} g(t) dt .$$

Then π_{t_1,t_2} is a continuous linear functional on H, and hence

$$\pi_{t_1,t_2}\left(\int_{R^n} K(x,y)f(y)dy\right) = \int_{R^n} \pi_{t_1,t_2}(K(x,y)f(y))dy \quad .$$

On the right side we can pass to the limit $t_1 \uparrow t$, $t_2 \downarrow t$ for all $t \in R_+$, by the dominated convergence theorem (since $x \notin \text{supp } f$). Because $L^2(R_+, \frac{dt}{t}) \subseteq L^1_{loc}(R_+, dt)$, we can apply the Lebesgue differentiation theorem to take the limit on the left side for almost all $t \in R_+$. Hence (2) follows, and K is the kernel of $\tilde{T}$.

The next thing to show is that $\tilde{T}$ is bounded on $L^2(R^n, dx)$, and to do this, we shall use **Plancherel's** theorem.

Let $f \in C_c^\infty(R^n)$. We want to estimate

$$\int_0^\infty \int_{R^n} [h(\frac{x-y}{t})\frac{1}{t^n}f(y)]^2 \frac{dydt}{t} \quad .$$

By Plancherel's theorem, this is equal to

$$\int_0^\infty \int_{R^n} [\hat{h}(t\xi)\hat{f}(\xi)]^2 \frac{d\xi dt}{t}$$

$$= \int_{R^n} |\hat{f}(\xi)|^2 [\int_0^\infty |\hat{h}(t\xi)|^2 \frac{dt}{t}] d\xi \quad .$$

To estimate $\int_0^\infty |\hat{h}(t\xi)|^2 \frac{dt}{t}$, we may assume that $|\xi| = 1$. Identify R^{n-1} and the subspace of R^n that is orthogonal to ξ, so that every $y \in R^n$ can be written as $y = \alpha\xi + z$, where $z \in R^{n-1}$ and $\alpha = \langle y, \xi \rangle \in R$. Thus

$$\hat{h}(t\xi) = \int_{R^n} e^{-2\pi i t\langle \xi, y\rangle} h(y)dy$$

$$= \int_{-\infty}^\infty e^{-2\pi i t\alpha} [\int_{R^{n-1}} h(\alpha\xi + z)dz] d\alpha \quad .$$

Write $h_\xi(\alpha) = \int_{R^{n-1}} h(\alpha\xi + z)dz$, so that $\hat{h}(t\xi) = \hat{h}_\xi(t)$.

We are now going to prove that $\int_0^\infty |\hat{h}(t\xi)|^2 dt$ and $\int_0^\infty |\hat{h}(t\xi)|^2 \frac{dt}{t^2}$ are bounded, which implies the boundedness of $\int_0^\infty |\hat{h}(t\xi)|^2 \frac{dt}{t}$.

For the first integral, we have

$$\int_0^{\infty} |\hat{h}(t\xi)|^2 dt \leq \int_{-\infty}^{\infty} |\hat{h}(t\xi)|^2 dt = \int_{-\infty}^{\infty} |h_\xi|^2 dt \quad .$$

The condition $|h(x)| \leq \dfrac{C_h}{(1+|x|)^{n+1}}$ implies that $|h_\xi(t)| \leq \dfrac{C_h}{(1+|t|)^2}$, which is enough to control the first integral.

To estimate the second integral, we begin by observing that $F(u) = \int_{-\infty}^{u} h_\xi(t)dt$ defines a function in $L^2(\mathbb{R})$. Indeed, the conditions $\int_{\mathbb{R}^n} h(x)dx = 0$ and $|h_\xi(t)| \leq \dfrac{C_h}{(1+|t|)^2}$ imply that $uF(u)$ is bounded; hence

$$C_h \geq \int_{-\infty}^{\infty} |F(u)|^2 du = C_h \int_{-\infty}^{\infty} |\hat{F}(u)|^2 du$$

$$= C_h \int_{-\infty}^{\infty} |\hat{h}_\xi(u)|^2 \frac{du}{u^2} \quad .$$

This proves that $\tilde{T}$ is bounded on L^2. Note that, if $|\hat{h}|$ is radial, then $\int_0^{\infty} |\hat{h}(t\xi)|^2 \frac{dt}{t}$ is a constant, and hence T is, up to a constant multiple, an isometry. This finishes the proof of the theorem.

For $R > 0$ fixed, the functions $h_{(v)}(x) = h(x-v)$, $|v| < R$, satisfy uniformly the conditions (i) and (ii) above. Furthermore, if $|\hat{h}|$ is radial, so is $|\widehat{h_{(v)}}|$ (and, in fact, $|\hat{h}| = |\widehat{h_{(v)}}|$). We can use this to construct a new Calderón-Zygmund operator as follows.

Let $B = B(0,R)$, let dv be Lebesgue measure on the ball B, and let $(A,\mu) = (\mathbb{R}_+, \frac{dt}{t}) \times (B, dv)$. Define

$$T'_{t,v} f(x) = ((h_{(v)})_t * f)(x) \; ,$$

and let H' denote the Hilbert space $L^2(A, d\mu)$. Then we can think of $T'_{t,u} f(x)$ as an element of H', for each x, and we can define T as before, so that $Tf(x) = \|T'f(x)\|_{H'}$.

<u>Theorem</u>: T' <u>is a</u> $C - H'$ CZO, <u>and if</u> $|\hat{h}|$ <u>is radial, then</u> T <u>is a Littlewood-Paley operator</u>.

<u>Proof</u>: The kernel K' of T' takes its values in $\mathcal{B}(\mathbb{C}, H')$, which we can identify with H', and we have

$$K'(x,y)(t,v) = (h_{(v)})_t(x-y) = \frac{1}{t^n} h\left(\frac{x-y}{t} - v\right) .$$

All of the properties of K' and T' follow from the corresponding properties of K and $\widetilde{T}$ above, by integrating over B. In particular, this theorem follows from the preceding one.

Two important examples of these operators are the G-function of Littlewood and Paley, and the S-function of Lusin. Let h be either

$$h_1 = C_n \frac{x}{(1+|x|^2)^{\frac{n+3}{2}}}$$

or

$$h_2 = C'_n \frac{|x|^2 - n}{(1+|x|^2)^{\frac{n+3}{2}}} .$$

(Note that h_1 takes its values in R^n.) The constants are chosen so that if $f \in L^2(R^n, dx)$, and if $u(x,t)$ is its Poisson integral, then

$$((h_1)_t * f)(x) = t(\nabla_x u)(x,t)$$

and

$$((h_2)_t * f)(x) = t(\frac{\partial}{\partial t} u)(x,t) .$$

Corresponding to these two choices for h, we have the two operators

$$G_1(f)(x) = [\int_0^\infty |\nabla_x u(x,t)|^2 t dt]^{\frac{1}{2}}$$

and

$$G_2(f)(x) = [\int_0^\infty |\frac{\partial}{\partial t} u(x,t)|^2 t dt]^{\frac{1}{2}} .$$

(Note that $f(x) = -\int_0^\infty \frac{\partial}{\partial t} u(x,t) dt$. The functions h for which $f(x) = \int_0^\infty h_t * f \frac{dt}{t}$ are easy to characterize, since this is, at least formally, equivalent to

$$\hat{f}(\xi) = \int_0^\infty \hat{h}(t\xi)\hat{f}(\xi) \frac{dt}{t} .$$

Thus, if $\int_0^\infty \hat{h}(t\xi) \frac{dt}{t} \equiv 1$, then $f(x) = \int_0^\infty h_t * f \frac{dt}{t}$, and conversely.)

To define the S-function, we let $h = h_1 \oplus h_2$ (which takes its values in R^{n+1}), so that $(h_t * f)(x) = t(\nabla u)(x,t)$. The corresponding Littlewood-Paley operator, in this case is given by

$$Sf(x) = \left[\int_0^\infty \int_B |T'_{t,v} f(x)|^2 \frac{dvdt}{t}\right]^{\frac{1}{2}}$$

$$= \left[\int_0^\infty \int_B |((h_{(v)})_t * f)(x)|^2 \frac{dvdt}{t}\right]^{\frac{1}{2}}$$

$$= \left[\int_0^\infty \int_B |\nabla u(x - vt, t)|^2 t dvdt\right]^{\frac{1}{2}}$$

$$= \left[\iint_{x+\Gamma} |\nabla u(y,t)|^2 t^{1-n} dydt\right]^{\frac{1}{2}}$$

where $\Gamma = \{(y,t) \in R_+^{n+1} : |y| \leqq Rt\}$.

See [SW] for more details, and see p.130 for an application of the weighted converse inequalities for the S-function.

III. The Littlewood-Paley G-function and Carleson Measures.

Definition: A measure μ on R_+^{n+1} is a Carleson measure if there is a positive constant C_μ such that for each cube Q on R_+^{n+1} that has a face $\pi(Q)$ lying in $\{0\} \times R^n$,

$$\mu(Q) \leqq C_\mu |\pi(Q)| .$$

If we equip the upper half plane R_+^2 with polar co-ordinates (r,θ), then $drd\theta$ gives an example of a Carleson measure.

Carleson measures can be shown to be precisely those measures μ for which the mapping $f \to u$, which sends a function f on R^n to its Poisson integral on R_+^{n+1}, is a bounded mapping from $L^p(R^n, dx)$ to $L^p(R_+^{n+1}, d\mu)$, $1 < p < \infty$ (see [FS]).

There is a more general version of this. Suppose $f \in L^1_{loc}(R_+^{n+1}, d\mu)$ and define $\mathfrak{M}f$ on R^n by

$$\mathfrak{M}f(x_0) = \sup_{|x - x_0| < t} |f(x,t)| .$$

Then we have the following:

Proposition: For all $\lambda > 0$ and $f \in L^1_{loc}(\mathbb{R}^{n+1}_+, d\mu)$,

$$\mu(\{(x,t) \in \mathbb{R}^{n+1}_+ : f(x,t) > \lambda\}) \leq C_\mu C_n |\{x \in \mathbb{R}^n : \mathfrak{m}f(x) > \lambda\}| .$$

An immediate consequence of this is

$$[\int_{\mathbb{R}^{n+1}_+} |f(x,t)|^p d\mu]^{\frac{1}{p}} \leq C_\mu C_n [\int_{\mathbb{R}^n} |\mathfrak{m}f(x)|^p dx]^{\frac{1}{p}}$$

for all $p > 0$. In particular, if $f(x,t) = \varphi_t * f(x)$, where φ is dominated by a decreasing radial function, then

$$[\int_{\mathbb{R}^{n+1}_+} |f(x,t)|^p d\mu]^{\frac{1}{p}} \leq C[\int_{\mathbb{R}^n} |f^*(x)|^p dx]^{\frac{1}{p}} .$$

We can also give a weighted version of the definition of Carleson measures. Suppose that ω is a weight on $\mathbb{R}^n$ and μ is a measure on $\mathbb{R}^{n+1}_+$. Then μ is a Carleson measure with respect to ω if

$$\int_Q \omega(x) d\mu(x,t) \leq C_\mu \int_{\pi(Q)} \omega(x) dx$$

for some $C_\mu > 0$ and for all cubes Q in $\mathbb{R}^{n+1}_+$ which have a face $\pi(Q)$ lying in $\{0\} \times \mathbb{R}^n$. If ωdx is a doubling measure, then the above proposition is still true (with only minor changes needed for its proof).

An important example of a Carleson measure, which is used in [FS], [CM2], and [CMcIM] is the following. We assume that h is a function on $\mathbb{R}^n$ satisfying $\int_{\mathbb{R}^n} h dx = 0$, $|h(x)| \leq \dfrac{C}{(1+|x|)^{n+1}}$, and $|\nabla h(x)| \leq \dfrac{C}{(1+|x|)^{n+2}}$

Proposition: For $a \in BMO(\mathbb{R}^n)$,

$$[(h_t * a)(x)]^2 \frac{dxdt}{t}$$

is a Carleson measure with respect to any weight ω in A_2.

Proof: Note that $h_t * a$ is well defined, since $\int_{\mathbb{R}^n} h dx = 0$, and by the growth assumption on $h(x)$. Let Q be a cube in $\mathbb{R}^{n+1}_+$ with a face $\pi(Q)$ lying in $\{0\} \times \mathbb{R}^n$, and we assume (without loss in generality) that $m_{\widetilde{\pi(Q)}} a = 0$. Let $a_1 = a\chi_{\widetilde{\pi(Q)}}$ and

$a_2 = a - a_1$. Then it is enough to show that

$$\int_Q (h_t * a_i)^2 \omega(x) \frac{dxdt}{t} \leq C(h,\omega)\|a\|_*^2 \int_{\pi(Q)} \omega(x)dx$$

for $i = 1,2$.

When $i = 1$ we use the fact that the Littlewood-Paley G-function is bounded on $L^2(\mathbb{R}^n, \omega dx)$ for $\omega \in A_2$ (since the G-function comes from a vector-valued CZO). Hence,

$$\int_Q |h_t * a_1|^2 \omega(x) \frac{dxdt}{t} = \int_{R^n} [\int_0^\infty |h_t * a_1|^2 \frac{dt}{t}]^{\frac{1}{2} \cdot 2} \omega(x)dx$$

$$\leq C(h,\omega) \int_{R^n} |a_1|^2 \omega(x)dx$$

$$= C(h,\omega) \int_{R^n} |a_1 - m_{\overline{\pi(Q)}}(a_1)|^2 \omega(x)dx \quad .$$

Let $\epsilon = \epsilon(\omega) > 0$ be such that ω satisfies the reverse Hölder inequality with exponent $1 + \epsilon$. Then also using Hölder's inequality, we obtain

$$\int_Q |h_t * a_1|^2 \omega(x) \frac{dxdt}{t}$$

$$\leq C(h,\omega) [\int_{\overline{\pi(Q)}} |a_1 - m_{\overline{\pi(Q)}}(a_1)|^{2\frac{1+\epsilon}{\epsilon}}]^{\frac{\epsilon}{1+\epsilon}} [\int_{\overline{\pi(Q)}} \omega^{1+\epsilon} dx]^{\frac{1}{1+\epsilon}} \quad .$$

If we use the equivalence on BMO of the norms $\|f\|_*$ and $\|f\|_{*,p}$, where $p = \frac{2(1+\epsilon)}{\epsilon}$

as well as the reverse Hölder condition on ω, we can conclude that

$$\int_Q |h_t * a_1|^2 \omega(x) \frac{dxdt}{t} \leq C(h,\omega)\|a\|_*^2 \int_{\overline{\pi(Q)}} \omega(x)dx \quad .$$

The desired estimate now follows from the fact that ωdx is a doubling measure on $\mathbb{R}^n$.

For $i = 2$ we use an earlier lemma regarding the growth of BMO functions. Let $\delta = \delta(Q)$ be the side length of Q, so that $Q = \pi(Q) \times [0,\delta]$. Then

$$\int_{\pi(Q)} \int_0^\delta |h_t * a_2|^2 \omega(x) \frac{dxdt}{t} = \int_{\pi(Q)} \int_0^\delta [\int_{R^n} \frac{1}{t^n} h(\frac{x-y}{t}) a_2(y)dy]^2 \omega(x) \frac{dxdt}{t}$$

$$\leq \int_{\pi(Q)} \int_0^\delta [\int_{R^n} C_h \frac{t|a_2(y)|dy}{t^{n+1} + |x-y|^{n+1}}]^2 \omega(x) \frac{dxdt}{t} \quad .$$

Since $m_{\overline{\pi(Q)}}a = 0$, we may apply the aforementioned growth lemma, and, also, the fact that a_2 is supported in $\overline{\pi(Q)}^c$, to obtain the following majorant of the previous expression:

$$\int_{\pi(Q)} \int_0^\delta [C_h t \frac{\|a\|_*}{\delta}]^2 \omega(x) \frac{dxdt}{t} \leq C_h \|a\|_*^2 (\int_{\pi(Q)} \omega(x)dx)(\int_0^\delta \frac{tdt}{\delta^2})$$

$$\leq C_h \|a\|_*^2 \int_{\pi(Q)} \omega(x)dx \quad .$$

But this is all we need.

Let us show how these weighted Carleson measures can be useful for studying certain operators. We shall do this in conjection with the extrapolation theorem of Rubio de Francia, stated at the end of Chapter 2.

From the facts about Carleson measures that we have mentioned so far, (which are still true in the weighted case with minor changes), we see that for $f \in L^2(\omega dx)$, $\omega \in A_2$, and φ a function on R^n dominated by an integrable decreasing radial function,

$$\int_{R_+^{n+1}} |\varphi_t * f|^2 |h_t * a|^2 \omega(x) \frac{dxdt}{t} \leq C \int_{R^n} |f^*|^2 \omega(x)dx$$

$$\leq C \int_{R^n} |f|^2 \omega(x)dx \quad .$$

Hence, the operator

$$f \to [\int_0^\infty |\varphi_t * f|^2 |h_t * a|^2 \frac{dt}{t}]^{\frac{1}{2}}$$

is bounded on $L^2(R^n, \omega dx)$ for any $\omega \in A_2$; thus, by the extrapolation theorem of Rubio de Francia, it is bounded on $L^p(\omega dx)$ for any $1 < p < \infty$ and any $\omega \in A_p$.

This operator is strongly related to the operator

$$f \to \int_0^\infty (\varphi_t * f)(h_t * a) \frac{dt}{t} ,$$

which is studied by Coifman and Meyer in [CM2], where they show its L^2-boundedness. The remarks above together with simple changes in their work show that this operator is also bounded on $L^p(\omega dx)$ for any $1 < p < \infty$ and any $\omega \in A_p$. This is not surprising, since this operator is mainly used to prove L^2-boundedness of operators associated with standard kernels and, thus, considered as candidates for CZO's .

IV. The Littlewood-Paley Dyadic Decomposition.

Another important type of Littlewood-Paley operator is obtained in the following way. Let $(E_j)_{j\in J}$ be a partition of R^n into countably many measurable subsets. Let $S_{E_j} f = (\hat{f}\chi_{E_j})^{v}$, and define T by

$$Tf = \left(\sum_{j\in J} |S_{E_j} f|^2\right)^{\frac{1}{2}} .$$

An easy application of Plancherel's theorem shows that this is a Littlewood-Paley operator (where (A,μ) is a discrete space). Of course, we are mainly interested in partitions $(E_j)_{j\in J}$ such that the assoicated operator T is bounded for some $p \neq 2$.

The most famous example is when $J = \{-1, +1\} \times Z$, $n = 1$, and for $j = (\epsilon, k)$ $E_j = \{\epsilon[2^k, 2^{k+1})\}$. (Here $\epsilon[a,b)$ means $[a,b)$ if $\epsilon = +1$ and $(-b,-a]$ if $\epsilon = -1$). The fact that the operator T associated to this partition is bounded on $L^p(dx)$ for $1 < p < \infty$ (and even on $L^p(\omega dx), \omega \in A_p$) can be seen as follows.

By the multiplier theorem of the preceding chapter, the symbols χ_{R_+} and χ_{R_-} define two CZO's, S_+ and S_-. As in the last chapter, we let $\varphi \in C_c^\infty(R)$ be a function equal to 1 on $[\frac{1}{2}, 1]$ and equal to 0 outside of $[\frac{1}{3}, \frac{4}{3}]$. For $j = (\epsilon, k) \in J$, let $\varphi_j(\cdot) = \varphi(\epsilon 2^{-k}\cdot)$ and let $f_j = (\hat{f}\varphi_j)^{v}$. Then the Hilbert space valued version of the multiplier theorem of the last chapter implies that

$$\left\|\left(\sum_{j\in J} |f_j|^2\right)^{\frac{1}{2}}\right\|_p \leq C_p(\varphi)\|f\|_p .$$

Observe that each E_j is the intersection of two infinite intervals which can be obtained from R_+ and R_- by translation, respectively. Thus

$$S_{E_j} f = M_1 S_+ M_2 S_- M_3 f_j ,$$

where M_1, M_2, and M_3 are multiplication operators corresponding to three unimodular functions (depending on j), which arise from taking the Fourier transform of translations. Therefore, if we apply Corollary 1 in Section III of Chapter 0 (see p 9) twice, we obtain

$$\left\|\left(\sum_{j\in J} |S_{E_j} f|^2\right)^{\frac{1}{2}}\right\|_p \leq C_p \left\|\left(\sum_{j\in J} |f_j|^2\right)^{\frac{1}{2}}\right\|_p \leq C(p,\varphi)\|f\|_p .$$

This result leads itself to an n-dimensional analogue. The partition on R^n that we take is simply the one we have used on R "raised to the power n" (in the appropriate way). Then the passage from the one-dimensional case to the n-dimensional case follows easily from Corollary 2 in Section III of Chapter 0 (p. 9) and the following "restriction" argument.

Let E be a subset of R and let $\tilde{E} \subseteq R^n$ be defined by $\tilde{E} = E \times R^{n-1}$. For a function f on R^n, we let $f_{x_2,\dots,x_n}$ be the function on R defined by

$$f_{x_2,\dots,x_n}(t) = f(t, x_2, \dots, x_n) ,$$

with $x_2,\dots,x_n$ being fixed. Then

$$S_E(f_{x_2,\dots,x_n}) = (S_{\tilde{E}} f)_{x_2,\dots,x_n} .$$

See [S] for details, as well as for a proof of the following consequence of these results, which we state in a special case.

<u>The Marcinkiewicz Multiplier Theorem</u>: <u>Suppose that</u> $\sigma : R^n \to R$ <u>satisfies the following inequalities</u>: <u>for any</u> $\alpha = (\alpha_1, \dots, \alpha_n)$, $\alpha_i \in \{0,1\}$,

$$\left| \frac{\partial^\alpha}{\partial \xi^\alpha} \sigma \right| \leqq C_\alpha \prod_{i=1}^{n} |\xi_i|^{-\alpha_i} .$$

<u>Then the operator defined for</u> $f \in \mathcal{S}(R^n)$ <u>by</u> $Tf = (\sigma \hat{f})^\vee$ <u>extends to a bounded operator on</u> $L^p(R^n, dx)$ <u>for</u> $1 < p < \infty$.

Note that this theorem, unlike the theorem of the preceding chapter, is not "isotropic".

Let us see now how this result may be used to prove the L^p-boundedness of a Littlewood-Paley operator in two dimensions, of which there is no one-dimensional analogue. Let $(\theta_j)_{j \in N}$ be a decreasing sequence such that $\theta_1 = 2\pi$ and $\lim_{n\to\infty} \theta_n = 0$, and let

$$E_j = \{ (x,y) \in R^2 : \theta_{j+1} \leqq \mathrm{Arg}(x+iy) < \theta_j \} .$$

Then we have the following.

<u>Theorem</u>: <u>If</u> (θ_j) <u>is a Lacunary sequence, then the Littlewood-Paley operator associated to</u> (E_j) <u>is bounded on</u> $L^p(R^2, dx)$, $1 < p < \infty$.

Proof: We shall use the maximal theorem of Nagel, Stein, and Wainger (see the end of Section III in Chapter 1 p. 9) and a technique of Cordoba and Fefferman [COF].

As in the previous proof, we shall first prove an inequality of the form

$$\|(\sum_{j=1}^{\infty} |f_j|^2)^{\frac{1}{2}}\|_p \leq C\|f\|_p ,$$

where the supports of the f_j's have finite overlap, and where $S_{E_j} f_j = S_{E_j} f$. To define the f_j , we shall need the following:

Lemma: Let $\varphi \in C_c^{\infty}(\mathbb{R}^n)$ be equal to 1 on $[-1,1]$ and 0 outside of $[-(1+\epsilon),1+\epsilon]$, where $\epsilon > 0$ is fixed. Let C be any fixed positive number. Then multipliers with symbol $\varphi(a \operatorname{Arg}(x+iy) - b)$, where $a>0$ and $b>0$, satisfy uniformly the hypotheses of the Marcinkiewicz multiplier theorem for all pairs (a,b) such that

$$\{(x,y) : \varphi(a \operatorname{Arg}(x+iy) - b) \neq 0\} \subsetneq \{(x,y) : ay \leq cx\} . \tag{1}$$

The proof of the lemma is straightforward; one simply computes the appropriate derivatives and estimates them using the condition $ay \leq cx$.

The geometric meaning of the condition (1) above is that the opening of the cone

$$\Gamma = \operatorname{supp} \varphi\, (a \operatorname{Arg}(x+iy) - b)$$

dominates its "distance" from the abscissa; that is, $\beta_2 \geq C\beta_1$ (see picture below).

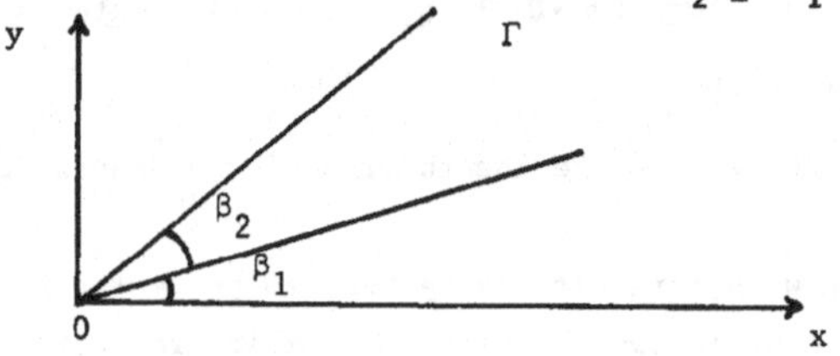

Choose positive numbers a_j, b_j such that

$$\{(x,y) : \theta_j \leq \operatorname{Arg}(x+iy) \leq \theta_{j+1}\}$$

$$= \{(x,y) : -1 \leq a_j \operatorname{Arg}(x+iy) - b_j \leq 1\} ,$$

and let Γ_j be the cone which corresponds to the pair (a_j, b_j) (and to φ) as above. Then, because $\{\theta_j\}$ is lacunary, there exists $C>0$ such that the pairs (a_j, b_j) satisfy the condition (1) above uniformly. Moreover, for $\epsilon > 0$ small enough, the cones $(1+\epsilon)\Gamma_j$ have bounded overlap.

Let $\varphi_j(x,y)=\varphi(a_j \operatorname{Arg}(x+iy)-b_j)$ for $j\in\mathbb{N}$. The lemma implies that the φ_j's are the symbols of multipliers which satisfy the hypotheses of the Marcinkiewicz multiplier theorem uniformly. Since their supports have bounded overlap, it now follows that the vector-valued symbol $(\varphi_j)_{j\in\mathbb{N}}$ satisfies the hypotheses of the vector-valued analogue of the Marcinkiewicz multiplier theorem. Thus, if we define f_j by $\hat{f}_j=\varphi_j\hat{f}$, then

$$\|(\sum_{j=1}^{\infty}|f_j|^2)^{\frac{1}{2}}\|_p \leq C_p\|f\|_p .$$

Note that $S_{E_j}f_j=S_{E_j}f$, by our choice of a_j's and b_j's.

Each cone E_j is the intersection of two half planes H_j and H_j', with $H_j' = H_{j+1}^c$. Thus $S_{E_j}=S_{H_j}S_{H_j'}$. The restriction argument that we mentioned at the end of the previous proof shows that S_{H_j} is bounded on $L^p(\omega dx)$, provided that ω is uniformly in A_p on the lines perpendicular to the boundary of H_j (S_{H_j} acts like the identity operator in the direction of the boundary of H_j, and it acts like a linear combination of the Hilbert transform and the identity in the perpendicular direction). If we let m_j denote the "partial" Hardy-Littlewood maximal function along this (perpendicular) direction, then we obtain a large collection of such weights by setting

$$\omega = [m_j(f^s)]^{\frac{1}{s}}$$

for $f\in L^s_{loc}(\mathbb{R}^2)$, where $s>1$. (This is because $(g^*)^\delta$ lies in A_1 uniformly for $g\in L^1_{loc}(\mathbb{R})$ and some fixed $\delta<1$, by the theorem of Coifman and Rochberg cited in Section II of Chapter 2.)

Suppose now that $f,g\in C_c^\infty(\mathbb{R}^2)$. Then for all $j\in\mathbb{N}$,

$$\int_{\mathbb{R}^2}|S_{E_j}f|^2 g\,dx = \int_{\mathbb{R}^2}|S_{H_j}S_{H_j'}f|^2 g\,dx$$

$$\leq C\int_{\mathbb{R}^2}|S_{H_j'}f|^2[m_j(g^s)]^{\frac{1}{s}}dx$$

$$\leq C\int_{\mathbb{R}^2}|f|^2[m_{j+1}(m_j(g^s))]^{\frac{1}{s}}dx$$

$$\leq C\int_{R^2}|f|^2[M_\theta(M_\theta(g^s))]^{\frac{1}{s}}dx \quad ,$$

where $M_\theta g(x) = \sup_{j\in N} m_j g(x)$, so that M_θ is bounded on $L^p(dx)$ for $1<p<\infty$, by the theorem of Nagel, Stein, and Wainger.

Finally, we must estimate $\|(\sum_j |S_j f|^2)^{\frac{1}{2}}\|_p$.

$$\|(\sum_j |S_j f|^2)^{\frac{1}{2}}\|_p = \|\sum_j |S_j f|^2\|_{\frac{p}{2}}^{\frac{1}{2}}$$

$$= (\sup_{\|g\|_{(\frac{p}{2})'}=1} \int_{R^2} \sum_{j=1}^{\infty} |S_j f|^2 g\, dx)^{\frac{1}{2}}$$

$$= (\sup_{\|g\|_{(\frac{p}{2})'}=1} \int_{R^2} \sum_{j=1}^{\infty} |S_j f_j|^2 g\, dx)^{\frac{1}{2}}$$

$$\leq (\sup_{\|g\|_{(\frac{p}{2})'}=1} \int_{R^2} \sum_j |f_j|^2 [M_\theta(M_\theta)(g^s))]^{\frac{1}{s}}dx)^{\frac{1}{2}}$$

$$\leq C_s\|\sum_j |f_j|^2\|_{\frac{p}{2}}^{\frac{1}{2}}$$

$$\leq C_s C_p\|f\|_p \quad .$$

Here C_s is the norm of the operator

$$g \to M_\theta(M_\theta(g^s))^{\frac{1}{s}}$$

on $L^{(\frac{p}{2})'}$, so that C_s is finite if $s<(\frac{p}{2})'$. This finishes the proof for $2\leq p<\infty$; the case of $1<p<2$ follows by duality.

For other applications of this "restricted weighted boundedness", and for more details on the relation between maximal operator results and Fourier multiplier problems, see [CO], [NSW] and [COF].

CHAPTER 7

THE CAUCHY KERNEL ON LIPSCHITZ CURVES

The Cauchy kernel has a long history. One of the first problems in which it appears is the following. Suppose that Γ is a smooth Jordan curve, and let Ω_+ and Ω_- be the two connected components of $\mathbb{C}\setminus\Gamma$. Then given a "nice" complex function f on Γ, we want to find two functions F_+ and F_- which are analytic in Ω_+ and Ω_-, respectively, and which extend continuously to Γ in such a way that

$$F_+|_\Gamma - F_-|_\Gamma = f .$$

This problem is discussed in [MU]. Plemelj gave a solution to this problem in [P], where he considers

$$F_+(z) = \frac{1}{2\pi i}\int_\Gamma f(t)\,\frac{dt}{(t-z)} \qquad \text{for} \quad z\in\Omega_+$$

and

$$F_-(z) = \frac{1}{2\pi i}\int_\Gamma f(t)\,\frac{dt}{(t-z)} \qquad \text{for} \quad z\in\Omega_- .$$

If f is, say, Hölder continuous and compactly supported, then F_+ and F_- have Hölder continuous boundary values on Γ, and

$$F_+|_\Gamma = \frac{1}{2}C(f) + \frac{1}{2}f$$

$$F_-|_\Gamma = \frac{1}{2}C(f) - \frac{1}{2}f ,$$

where

$$C(f) = \frac{1}{2\pi i}\int_\Gamma \frac{f(t)}{(t-s)}\,dt .$$

(This integral is taken in the principal value sense, and clearly exists at any point where f is differentiable).

The Cauchy kernel associated to a Jordan curve Γ is then defined on $\mathbb{R}\times\mathbb{R}\setminus\Delta$ by

$$K_\Gamma(x,y) = \frac{1}{\pi i}\,\frac{1}{z(x)-z(y)} ,$$

where $z(\cdot)$ is an arc length parameterization of Γ. In particular, if Γ is the real axis, then K_Γ is the kernel of the Hilbert transform. Notice that if Γ is the graph of a Lipschitz function φ, then

$$\frac{1}{|x-y|} \leqq \frac{1}{|z(x)-z(y)|} \leqq \frac{\sqrt{1+\|\varphi'\|_\infty^2}}{|x-y|},$$

from which one can easily show that K_Γ is a standard kernel.

Calderón's interest in the Cauchy kernel was not motivated by its connection with analytic functions, but by its connection with certain so-called commutators, which appear to be very important in the theory of partial differential equations. (See [C3], for instance.) These commutators are singular integral operators with kernels

$$\frac{1}{x-y}\left(\frac{A(x)-A(y)}{x-y}\right)^n,$$

where $n \in N$ and A is a real valued function on R satisfying $\|A'\|_\infty < \infty$. They are called commutators because of the following identity. For a linear operator acting on real valued functions on R, let $C_A(T) = TA - AT$, where A also denotes the operator of multiplication by the function A. Define $C_A^m(T)$ by iteration (of C_A), and let D denote the operator $\frac{d}{dx}$. Then, it can be shown that,

$$C_A^m(HD^m)f(x) = \int_{-\infty}^{\infty} \frac{(-1)^m m!}{(x-y)} \left[\frac{A(x)-A(y)}{x-y}\right]^m f(y)dy .$$

The connection between these commutators and the Cauchy kernel is given in some detail in the next chapter. (See also [C3].) What comes out is the following. If there exists $\delta > 0$ such that K_Γ is a C-Z kernel whenever Γ is the graph of a Lipschitz function φ satisfying $\|\varphi'\|_\infty < \delta$, then the above commutator kernels are actually C-Z kernels. A.P. Calderón proved that the first such commutator is a C-Z kernel, in 1965 [C1]. Then, in 1976, Coifman and Meyer showed that all of those commutators are C-Z kernels, but they did not obtain satisfactory estimates on the growth (in m) of the norms of those kernels. In 1977, an estimate of the form $C^m\|A'\|_\infty^m$ was obtained by Calderón as a corollary of the following result.

Theorem: (A.P. Calderón [C2].) There exists $\delta > 0$ such that if φ is a Lipschitz function satisfying $\|\varphi'\|_\infty < \delta$, and if Γ denotes the graph of φ, then K_Γ is a CZK with uniformly bounded norm.

In 1981, Coifman, Meyer, and MacIntosh proved directly that all the above commutators were CZK's , and they obtained the norm estimate $C\|A'\|_\infty^n(1+n)^4$ ([CMcIM]) . In particular, the result of A.P. Calderón may be recovered from this. Later, in 1981, Guy David developed a real variable technique to relax the condition $\|\varphi'\|_\infty < \delta$ to, simply, $\|\varphi'\|_\infty < \infty$, via a perturbation argument. (See [D].)

The proof of A.P. Calderón uses essentially the same techniques as the proof that he gave for the first commutator, as well as some results on the conformal mapping from Ω_+ to the upper half plan $\prod_+$. The variant of his proof which we present here is due to Yves Meyer, and it uses the original proof of Coifman and Meyer in 1976, whereas Calderón uses techniques from [Cl]. We chose the proof we shall present because it is unpublished, and also because it is somewhat more instructive for those who wish to master the Calderón-Zygmund machinery. As a corollary to the result of A.P. Calderón and its generalizations, we shall see that, if we restrict the type of convergence towards the curve Γ , one can answer the problem stated at the beginning of this chapter for much more general curves than smooth Jordan curves, and functions more general than those which are Hölder continuous and compactly supported.

We come now to the proofs. As we saw earlier (see the last section of Chapter 4), we can restrict ourselves to the case where $\varphi \in C_c^\infty(\mathbb{R})$; in fact, we make this a priori assumption throughout the following preliminary remarks.

In what follows, Γ denotes the graph of φ , that is, $\{(x,y) : y = \varphi(x)\}$, and we let $z(\cdot)$ be an arc length parameterization of Γ . For a function f on $\mathbb{R}$ we denote by $\tilde{f}$ the function induced on Γ by f ; that is, $\tilde{f}((x,\varphi(x)) = f(x)$.

(1) <u>Regularity of F_+ and F_- near Γ</u> .

Suppose that $f \in C_c^\infty(\mathbb{R})$, and let F_+ and F_- be associated to f as above. Then we have the following.

<u>Lemma</u>: <u>F_+ and F_- extend continuously to the boundary, and so do all of their derivatives. Furthermore,</u>

$$\lim_{\substack{z \to z(s) \\ z \in \Omega_+}} F_+(z) = \frac{1}{2}\tilde{f}(z(s)) + \frac{1}{2\pi i}\,\text{P.V.}\int \frac{\tilde{f}(z(u))z'(u)du}{z(u) - z(s)} ,$$

and

$$\lim_{\substack{z \to z(s) \\ z \in \Omega_-}} F_-(z) = -\frac{1}{2}\tilde{f}(z(s)) + \frac{1}{2\pi i}\,\text{P.V.}\int \frac{\tilde{f}(z(u))z'(u)du}{z(u) - z(s)} .$$

<u>Proof</u>: We consider only F_+ , the proof for F_- being similar. To prove the first assertion, we must show that F_+ and all of its derivatives are locally uniformly continuous near the boundary. To do this, it suffices to show that $F_+^{(n)}$ is

locally bounded near the boundary, for $n = 0, 1, 2, \ldots$. This is obvious for $n = 0$.

For $n = 1$, we integrate by parts to obtain

$$F_+^{(1)}(z(s)) = \frac{1}{2\pi i}\int_{\mathbb{R}} \frac{\tilde{f}(z(u))z'(u)du}{[z(u) - z(s)]^2}$$

$$= \frac{1}{2\pi i}\int_{\mathbb{R}} \frac{(\tilde{f}\circ z)'(u)du}{[z(u) - z(s)]} .$$

This reduces us to the case $n = 0$, since $\frac{(\tilde{f}\circ z)'}{z'}$ enjoys the same properties as $\tilde{f}\circ z$. The general case follows by induction.

To prove the second part of the lemma, we note that both the principal value integral and the above boundary limit exist everywhere. Hence, it will be enough for us to show that for any $\epsilon > 0$,

$$\left| \lim_{\substack{z\to z(s)\\ z\in\Omega_+}} F_+(z) - \frac{1}{2\pi i}\int_{|u-s|>\epsilon} \frac{\tilde{f}(z(u))z'(u)du}{z(u) - z(s)} - \frac{1}{2\pi i}\tilde{f}(z(s))\log\left(\frac{z(s+\epsilon) - z(s)}{z(s-\epsilon) - z(s)}\right)\right| \leq C\epsilon\|f'\|_\infty$$

We do this by computing $\lim_{\delta\to 0^+} F_+(z(s) + i\delta)$.

$$\lim_{\delta\to 0^+} F_+(z(s) + i\delta) = \frac{1}{2\pi i}\int_{|u-s|>\epsilon} \frac{\tilde{f}(z(u))z'(u)du}{z(u) - z(s)}$$

$$+ \lim_{\delta\to 0^+} \frac{1}{2\pi i}\int_{|u-s|<\epsilon} \frac{(\tilde{f}(z(u)) - \tilde{f}(z(s)))z'(u)du}{[z(u) - z(s) - i\delta]}$$

$$+ \lim_{\delta\to 0^+} \frac{1}{2\pi i}\int_{|u-s|<\epsilon} \frac{\tilde{f}(z(s))z'(u)du}{[z(u) - z(s) - i\delta]} .$$

Hence (1) follows immediately, and we conclude the proof of the lemma by letting ϵ tend to 0 .

(2) <u>Some properties of the conformal mapping from the upper half plane Π_+ onto Ω_+</u> .

Let $\Pi_+ = \{z\in\mathbb{C} : z = x + iy,\ y > 0\}$ and let $\Omega_+ = \{z\in\mathbb{C} : z = x + iy,\ y > \varphi(x)\}$. Consider the conformal mapping Φ from Π_+ onto Ω_+ which takes ∞ to itself, preserves the positive orientation, and which has the property that Φ and Φ' both extend continuously to the boundary. (We shall not consider the problem of the existence of such a conformal mapping; this is automatically ensured by the condition $\varphi \in C_c^\infty(R)$, and weaker conditions would do the job as well. Other such technical questions that we shall not deal with, since they are not part of our main interest, are whether or not a given function is the Poisson integral of its boundary values, where do the latter exist, etc. Indeed, we may do this. Unfortunately, no reference seems to be available to answer these questions without any work, so that we are forced to leave them as exercises.)

We are now going to prove some lemmas concerning this conformal mapping. The first will be used almost immediately, but the second one will not be needed until the next chapter.

<u>Lemma 1</u>: $|\Phi'| \in A_2$ and

$$|\Phi'(x+iy)| \approx m_{[x-y,x+y]}|\Phi'(t)| \quad .$$

<u>Proof</u>: The maximum principle applied to $\operatorname{Im}(\log \Phi')$ implies that this function is bounded by its supremum on R, which is $\arctan(\|\varphi'\|_\infty)$. Obvious geometric considerations imply now that $|\Phi'|$ is equivalent to $\operatorname{Re}\Phi'$, which is positive, and the same is also true for $\frac{1}{\Phi'}$. If we now use the fact that $\operatorname{Re}\Phi'$ is a harmonic function, and also the Poisson integral of its boundary values, we obtain that

$$\operatorname{Re}\Phi'(x+iy) = \frac{1}{2\pi}\int_{-\infty}^{\infty} \frac{y}{(x-t)^2+y^2} \operatorname{Re}\Phi'(t)dt$$

$$\geqq \frac{1}{2\pi}\int_{x-y}^{x+y} \frac{y}{(x-t)^2+y^2} \operatorname{Re}\Phi'(t)dt$$

$$\geqq C\, m_{[x-y,x+y]}(\operatorname{Re}\Phi') \quad .$$

Hence $|\Phi'(x+iy)| \geqq C\, m_{[x-y,x+y]}(|\Phi'|)$. Since the same is true for $\frac{1}{\Phi'}$, we obtain simultaneously that $|\Phi'|$ is an A_2 weight, and that the last inequality above is actually an equivalence, since

$$1 \leqq m_{[x-y,x+y]}(|\Phi'|)\, m_{[x-y,x+y]}\left(\frac{1}{|\Phi'|}\right) ,$$

by Schwartz's inequality. This concludes the proof of Lemma 1.

<u>Lemma 2</u>: <u>Let</u> $z = x+iy$, $z' = \Phi(z)$, <u>and suppose</u> $\eta > 0$ <u>is such that</u> $z' - i\eta \in \Gamma$.

Then

$$\eta \approx \int_{x-y}^{x+y} |\Phi'(t)|\,dt \quad .$$

<u>Proof</u>: Let L be the curve in Π_+ which is the inverse image of the vertical segment $[z'-i\eta, z']$ under Φ^{-1}. Since $|\mathrm{Arg}\,\Phi'| \leqq \arctan(\|\varphi'\|_\infty)$, as was noted in the preceding proof, and hence $|\mathrm{Arg}(\Phi^{-1})'| \leqq \arctan(\|\varphi^1\|_\infty)$, we conclude that L is the graph of a Lipschitz function ψ defined on $[0,y]$ which satisfies $\psi(y) = x$ and $\|\psi'\|_\infty \leqq \|\varphi'\|_\infty$.

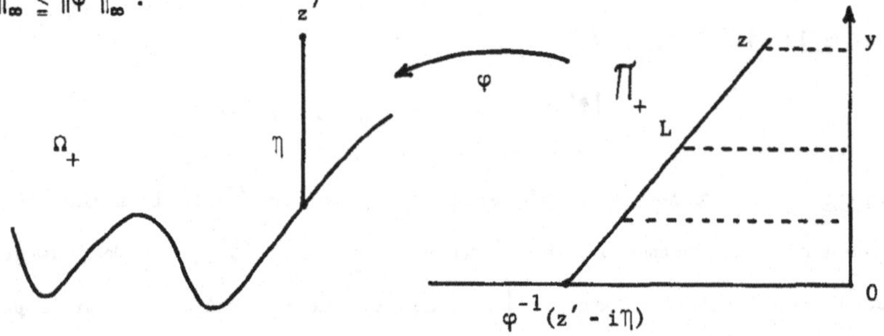

Parameterize L by $t \to (\psi(t), t)$, so that

$$\eta = \int_0^y |\Phi'(\psi(t)+it)| \sqrt{1+|\psi'(t)|^2}\,dt$$

$$\approx \int_0^y |\Phi'(\psi(t)+it)|\,dt \;,$$

since $\|\psi'\|_\infty \leqq \|\varphi'\|_\infty$. Hence, by Lemma 1

$$\eta \approx \int_0^y m_{[\psi(t)-t,\psi(t)+t]}(|\Phi'|)\,dt \;.$$

Using the fact that $|\Phi'|dx$ satisfies a doubling condition (since $|\Phi'|$ is a good weight, by the previous lemma), that $\|\psi'\|_\infty \leqq \|\varphi'\|_\infty$, and $|\psi(t)-x| = |\psi(t)-\psi(y)| \leqq Ct$ for $t > \frac{y}{2}$, we obtain

$$\eta \geqq C \int_{\frac{y}{2}}^{y} m_{[\psi(t)-t,\psi(t)+t]}(|\Phi'|)\,dt$$

$$\geqq C y\, m_{[x-y,x+y]}(|\Phi'|)$$

$$\approx \int_{x-y}^{x+y} |\Phi'(t)|\,dt \ .$$

On the other hand, since $|\Phi'| \in A_2 \subset A_\infty$, there exists $\alpha > 0$ such that (for $t \leqq y$)

$$m_{[\psi(t)-t,\psi(t)+t]}|\Phi'| \leqq C m_{[\psi(t)-y,\psi(t)+y]}(|\Phi'|)(\frac{t}{y})^{\alpha-1} .$$

Now $|\psi(t)-x| = |\psi(t)-\psi(y)| \leqq Cy$ for $t \in [0,y]$; hence,

$$m_{[\psi(t)-t,\psi(t)+t]}(|\Phi'|) \leqq C(\frac{t}{y})^{\alpha-1} m_{[x-y,x+y]}(|\Phi'|) ,$$

and we obtain the inequality

$$\eta \leqq C \int_{x-y}^{x+y} |\Phi'(t)|\,dt$$

as desired.

(3) <u>Boundedness of the Cauchy kernel in the regular case</u>.

We want to prove that the kernel

$$\frac{1+i\varphi'(y)}{x+i\varphi(x)-y-i\varphi(y)}$$

determines a bounded operator on L^2, where φ satisfies $\|\varphi'\|_\infty < \delta$. (This is equivalent to the L^2 boundedness of the Cauchy kernel, since $\|\varphi'\|_\infty < \infty$ implies that we may make the appropriate change of parameterization and that we can multiply the numerator by $1+i\varphi'(y)$.) As we noted earlier, it is enough to prove the result for $\varphi \in C_c^\infty(\mathbb{R})$, with estimates that depend only on $\|\varphi'\|_\infty$. In this case, it is clear that the principal value defined by the kernel exists.

To prove the boundedness of this kernel, we shall proceed as follows. Let $\lambda \in [0,1]$, and let T_0^λ be the principal value operator with kernel

$$\frac{1+i\lambda\varphi'(y)}{x+i\lambda\varphi(x)-y-i\lambda\varphi(y)}$$

Then we shall show that

$$\|\frac{d}{d\lambda} T_0^\lambda\|_{2,2} \leqq C(1+\|T_0^\lambda\|_{2,2}^2) \ .$$

Because T_0^λ is a multiple of the Hilbert transform if $\lambda = 0$, it follows from this differential inequality that

$$\|T_0^\lambda\| \leq y(\lambda),$$

where $y(\lambda)$ is the solution of

$$y' = C(1+y^2), \quad y(0) = C' > 0.$$

Thus $y(\lambda) = \tan(C\lambda + \arctan C')$, and $y(\lambda)$ is finite for all $\lambda < \delta$ if δ is small enough. This would imply that the above kernel is bounded on L^2 if $\|\varphi'\|_\infty$ is small enough.

It is convenient to know, a priori, that T_0^λ and $\frac{d}{d\lambda}T_0^\lambda$ are bounded on L^2, in order to be sure that the above differential inequality has a meaning. The regularity assumption on φ implies this easily, as we now prove for T_0^λ, the other one being similar.

<u>Lemma</u>: $(T_0^\lambda)_\epsilon$ <u>is uniformly bounded when</u> $\epsilon > 0$ <u>and</u> $\lambda \in [0,1]$.

<u>Proof</u>: Let $f \in C_c^\infty(\mathbb{R})$. Then

$$(T_0^\lambda)_\epsilon f(t) = \frac{1}{2\pi i} \int_{|s-t|>\epsilon} \frac{f(s)(1+i\lambda\varphi'(s))ds}{(s-t)+i\lambda(\varphi(s)-\varphi(t))}.$$

Because $\|\varphi'\|_\infty < \infty$, we may incorporate $(1+i\lambda\varphi')$ into f. The idea now is to write the Cauchy kernel as a perturbation of the Hilbert transform, and use Fourier analysis to control the perturbation.

$$\frac{1}{(s-t)+i\lambda(\varphi(s)-\varphi(t))} = \frac{1}{s-t}\,\frac{1}{1+i\lambda\left(\frac{\varphi(s)-\varphi(t)}{s-t}\right)}$$

$$= \frac{1}{s-t}\left[1 - \frac{i\lambda\left(\frac{\varphi(s)-\varphi(t)}{s-t}\right)}{1+i\lambda\left(\frac{\varphi(s)-\varphi(t)}{s-t}\right)}\right].$$

Rewrite this expression in the form

$$\frac{1}{s-t}\,[1+k_\lambda(s,t)].$$

It is easy to see that k_λ satisfies the following properties, uniformly in λ. Let $M > 0$ be such that $\operatorname{supp}\varphi \subseteq [-M,M]$. Then $k_\lambda(s,t) = 0$ if $|s| > M$ and $|t| > M$, and if $|s| < M$ and $|t| > 2M$, then for any $m, n \in N$,

$$\left| \frac{\partial^{m+n}}{\partial s^m \partial t^n} k_\lambda(s,t) \right| \leqq \frac{C_{m,n}}{|t|} .$$

Apply the (two-dimensional) inverse Fourier transform formula to k_λ to obtain

$$(T_0^\lambda)_\epsilon f(t) = \frac{1}{2\pi i} \int_{|s-t|>\epsilon} f(s) \frac{1}{s-t} \left(1 + \int_{R^2} \hat{k}_\lambda(u,v) e^{i(us+vt)} du dv\right) ds .$$

Suppose now that $\hat{k}_\lambda$ is in $L^1(R^2)$. Then we can interchange the order of integrations, and the above expression becomes

$$(T_0^\lambda)_\epsilon f(t) = \frac{1}{2\pi i} \int_{|s-t|>\epsilon} f(s) \frac{1}{s-t} ds$$

$$+ \frac{1}{2\pi i} \int_{R^2} e^{ivt} \hat{k}_\lambda(u,v) \left[\int_{|s-t|>\epsilon} \frac{f(s) e^{ius}}{s-t} ds \right] du dv .$$

Now using Minkowski's integral inequality and the L^2 boundedness of H_*, where

$$H_* f(t) = \sup_{\epsilon>0} \left| \frac{1}{\pi i} \int_{|s-t|>\epsilon} \frac{f(s)}{s-t} ds \right| ,$$

we obtain

$$\|(T_0^\lambda)_\epsilon\|_{2,2} \leqq C + C\|\hat{k}_\lambda\|_1 .$$

Note that this argument shows that $(T_0^\lambda)_*$ is bounded on L^2.

To show that $\hat{k}_\lambda$ is in $L^1(R^2)$, we use the properties of k_λ above, which imply that

$$\int_{R^2} |\hat{k}_\lambda| du dv \leqq \left[\int_{R^2} \left(\frac{1}{1+u^2+v^2} \right)^2 du dv \right]^{\frac{1}{2}} \left[\int_{R^2} \left((1+u^2+v^2) |\hat{k}_\lambda| \right)^2 du dv \right]^{\frac{1}{2}} \leqq C ,$$

where the latter integral is shown to be bounded by an application of Plancherel's theorem.

Let us now prove the theorem. We begin by computing $\frac{d}{d\lambda} T_0^\lambda$.

$$\frac{d}{d\lambda}[(T_0^\lambda)_\epsilon f(t)] = \frac{1}{2\pi i} \int_{|s-t|>\epsilon} \frac{f(s)\varphi'(s) ds}{(s-t) + i\lambda(\varphi(s) - \varphi(t))}$$

$$- \frac{1}{2\pi i} \int_{|s-t|>\epsilon} \frac{f(s)i[\varphi(s)-\varphi(t)](1+i\lambda\varphi'(s))ds}{[(s-t)+i\lambda(\varphi(s)-\varphi(t))]^2} .$$

For convenience, we shall write $z_\lambda(s) = s + i\lambda\varphi(s)$. These two integrands are of the form $\frac{1}{s-t}\psi_\lambda(s,t)$, where $\psi_\lambda(s,t)$ lies in $C^\infty(\mathbb{R}^2)$ and vanishes when s and t are both large enough. Thus the convergence of these integrals as ϵ tends to 0 is uniform in ϵ for t restricted to any fixed compact set; hence, the inequality that we seek will be obtained by passing to the limit.

Observe that $f(s)[\varphi(s)-\varphi(t)]$ can also be written as

$$f(s)\varphi(s) - f(t)\varphi(t) - [f(s)-f(t)]\varphi(t) .$$

Hence, the second term above is a commutator. To be precise, let φ denote also the the operator of multiplication by φ, let M_λ denote the operator of multiplication by $\frac{\varphi'}{1+i\lambda\varphi'}$, and define T_1^λ by

$$T_1^\lambda f(t) = \frac{1}{2\pi i} \text{ P.V. } \int \frac{[f(s)-f(t)][1+i\lambda\varphi'(s)]ds}{[(s-t)+i\lambda(\varphi(s)-\varphi(t))]^2} .$$

Then the previous formula becomes

$$\frac{\partial}{\partial\lambda} T_0^\lambda = T_0^\lambda M_\lambda - i(T_1^\lambda\varphi - \varphi T_1^\lambda) .$$

Thus the differential inequality that we want will follow from

$$\|T_1^\lambda\varphi - \varphi T_1^\lambda\|_{2,2} \leq C(1+\|T_0^\lambda\|_{2,2}^2) .$$

It is easy to check that $T_1^\lambda\varphi - \varphi T_1^\lambda$ is bounded on L^2, just like we did for T_0^λ. On the other hand, the kernels of $T_1^\lambda\varphi - \varphi T_1^\lambda$ and T_0^λ are, to within the multiplicative factor $(1+i\lambda\varphi'(s))$, standard kernels with uniformly bounded constants for $\lambda \in [0,1]$ and $\|\varphi'\|_\infty < \delta$. Hence it follows from the results of Chapter 4 that the inequality above will be a consequence of

$$(1) \qquad \|T_1^\lambda\varphi - \varphi T_1^\lambda\|_{5,5} \leq C(1+\|T_0^\lambda\|_{4,4}^2) .$$

We may assume that $\lambda = 1$, and, accordingly, we can drop the λ-superscript altogether.

The remainder of the proof of the theorem will be divided into two parts. First we shall use a succession of "miracles" to reduce the proof of (1) to the following problem, which we shall consider in the second part.

Let H_+, F_+, and G_+ be three holomorphic functions in Ω_+ which extend smoothly to the boundary Γ_φ of Ω_+, and assume that they satisfy Calderón's equation

$$H_+' = F_+ G_+' .$$

Suppose, further, that $|F_+| = O(\frac{1}{|z|})$ at ∞ and $|G_+'| = O(\frac{1}{|z|})$ at ∞. Then H_+ can be chosen so that $\lim_{|z|\to\infty} H_+(z) = 0$, and with this determination of H_+,

$$\|H_+\|_{L^2(\Gamma,ds)} \leq C\|F_+\|_{L^4(\Gamma,ds)} \|G_+\|_{L^4(\Gamma,ds)} .$$

Part 1: Reduction to the Calderón's Equation.

To prove (1) above from the Calderón's equation, we first reduce (1) to the following:

(2) $$\|T_1(\varphi f) - \varphi T_1(f)\|_{\frac{4}{3}} \leq C\|\varphi'\|_2 \|f\|_4 (\|T_0\|_{4,4}^2 + 1).$$

Indeed, to obtain (1) from (2), we shall prove that (2) implies

(3) $$(T_1(\varphi f) - \varphi T_1(f))^{\#} \leq C(1 + \|T_0\|_{4,4}^2)((f^4)^*)^{\frac{1}{4}} .$$

We know a priori that if f lies in $L^5(\mathbb{R},dx)$, then $[T_1(\varphi f) - \varphi T_1(f)] \in L^5$; moreover, since $f \to ((f^4)^*)^{\frac{1}{4}}$ is bounded on L^5, it is clear that (3) implies (1). (Note that we have used a result from Section III of Chapter 3 to control a function g by $g^{\#}$.)

To prove (3), pick $x_0 \in \mathbb{R}$ and let I be an interval containing x_0. Choose $\psi \in C_c^\infty(\mathbb{R})$ so that $\psi \equiv 0$ outside of $4I$, $\psi(x) = \varphi(x)$ for $x \in 2I$, and $\|\psi'\|_\infty \leq 2\|\varphi'\|_\infty \leq 2\delta$.

Let $f_1 = f\chi_{2I}$ and $f_2 = f - f_1$. Because the kernel of $T_1\varphi - \varphi T_1$ is, within a multiplication by the y-variable, a standard kernel, we have

$$|T_1(\varphi f_2)(x) - (\varphi T_1 f_2)(x) - T(\varphi f_2)(x_0) - (\varphi T_1 f_2)(x_0)| \leq C f^*(x_0)$$

for $x \in I$. Hence, to prove (3), it is enough to estimate

$$\frac{1}{|I|}\int_I |(T_1\varphi - \varphi T_1)f_1|\,dx ,$$

This is clearly equal to

$$\frac{1}{|I|}\int_I |(T_1\psi - \psi T_1)f_1|\,dx .$$

By (2) and Hölder's inequality, this is dominated by

$$\frac{1}{|I|^{\frac{3}{4}}}\left[\int_I |T_1\psi - \psi T_1)f|^{\frac{4}{3}}dx\right]^{\frac{3}{4}}$$

$$\leq \frac{C}{|I|^{\frac{3}{4}}}\|f_1\|_4\|\psi'\|_2(1+\|T_0\|^2_{4,4})$$

$$\leq C((f^4)^*)^{\frac{1}{4}}(x_0)\|\psi'\|_\infty(1+\|T_0\|^2_{4,4}) .$$

To reduce (2) to the Calderón's equation, we need the following identity. Let $\langle f|g\rangle_{d\mu} = \int fg\,d\mu$.

__Lemma__: __For__ $f,g \in C_c^\infty(\mathbb{R})$,

$$\langle f|T_1g\rangle_{(1+i\varphi')dx} = \langle g|T_1f\rangle_{(1+i\varphi')dx} .$$

To prove the lemma, observe that

$$\langle f|T_1g\rangle_{(1+i\varphi')dx} = \lim_{\epsilon\to 0}\iint_{|s-t|>\epsilon}\frac{f(t)[g(s)-g(t)](1+i\varphi'(s))(1+i\varphi'(t))}{[(s-t)+i(\varphi(s)-\varphi(t))]^2}\,ds\,dt .$$

Because of the symmetry properties of this expression, $f(t)g(s)$ can be replaced by $f(s)g(t)$, so that $f(t)[g(s)-g(t)]$ can be replaced by $[f(s)-f(t)]g(t)$. Hence f and g may be interchanged, which is what we wanted.

Let $f,g \in C_c^\infty(\mathbb{R})$, and apply the lemma to obtain that

$$\langle T_1(\varphi f) - \varphi T_1 f|g\rangle_{(1+i\varphi')dx} = \langle\varphi|fT_1(g) - gT_1(f)\rangle_{(1+i\varphi')dx}$$

$$= \langle\varphi|[fT_1(g) - gT_1(f)](1+i\varphi')\rangle dx$$

$$= -\langle \varphi' | \ell \rangle_{dx} ,$$

where ℓ is a primitive of

$$[fT_1(g) - gT_1(f)](1 + i\varphi') .$$

Hence,

$$\|T_1(\varphi f) - \varphi T_1(f)\|_{\frac{4}{3}} \leq C\|\varphi'\|_2 \sup_{\|g\|_4 = 1} \|\ell\|_2 .$$

Let F_+, F_-, G_+, and G_- be as before, so that $f(t) = F_+(t + i\varphi(t)) - F_-(t + i\varphi(t))$ and

$$F_+(t + i\varphi(t)) = \frac{1}{2} T_0 f(t) + \frac{1}{2} f(t) .$$

The reason why the Calderón's equation shows up in the estimate of $\|\ell\|_2$ comes from the following.

<u>Lemma</u>: $(T_1 f)(t) = \frac{1}{2} [F'_+(t + i\varphi(t)) + F'_-(t + i\varphi(t))]$.

Let $z(x) = x + i\varphi(x)$. Then to prove the lemma it is enough to show that

$$(T_1 f)(x) z'(x) = (T_0 f)'(x) .$$

Let $\epsilon > 0$, and let us compute $[(T_0)_\epsilon f]'(x)$.

$$\frac{1}{2\pi i} \frac{d}{dx} \int_{|y-x|>\epsilon} \frac{1}{z(y) - z(x)} f(y) dz(y) = \frac{1}{2\pi i} \int_{|y-x|>\epsilon} \frac{z'(x)}{[z(y) - z(x)]^2} f(y) dz(y)$$

$$+ \frac{1}{2\pi i} \frac{f(x-\epsilon) z'(x-\epsilon)}{z(x-\epsilon) - z(x)} - \frac{1}{2\pi i} \frac{f(x+\epsilon) z'(x+\epsilon)}{z(x+\epsilon) - z(x)}$$

$$= \frac{1}{2\pi i} \int_{|y-x|>\epsilon} \frac{z'(x)(f(y) - f(x))}{[z(y) - z(x)]^2} z'(y) dy$$

$$+ \frac{1}{2\pi i} \frac{(fz')(x-\epsilon) - (fz')(x)}{z(x-\epsilon) - z(x)}$$

$$- \frac{1}{2\pi i} \frac{(fz')(x+\epsilon) - (fz')(x)}{z(x+\epsilon) - z(x)} .$$

As usual, the regularity of the functions involved implies that convergence as ϵ tends to 0 is uniform, at least locally in x , from which the lemma follows easily.

Let us use this lemma to estimate $\|\ell\|_2$. Let $h = fT_1 g - gT_1 f$, so that the formula $f = F_+ \circ z - F_- \; z$ and the lemma imply that

$$h = \frac{1}{2}[F_+G'_+ + F_+G'_- - F_-G'_+ - F_-G'_- - F'_+G_+ + F'_+G_- - F'_-G_+ + F'_-G_-] \circ z$$

$$= \frac{1}{2}[2F_+G'_+ - 2F_-G'_- - (F_+G_+)' + (F_-G_-)' - (F_-G_+)' + (F_+G_-)] \circ z \, .$$

Define H_+ and H_- in terms of F_+, G_+, F_-, and G_- as in the Calderón's equation above. Then

$$h(x)z'(x) = \frac{d}{dx}[H_+ \circ z - H_- \circ z]$$

$$+ \frac{1}{2}\frac{d}{dx}([F_-G_- + F_+G_- - F_-G_+ - F_+G_+](z(x))) \; .$$

Thus

$$\langle \varphi | h \rangle_{dx} = -\langle \varphi' | \ell \rangle_{dx} \, ,$$

and, hence,

$$\|\ell\|_2 \leqq \|H_+\|_2 + \|H_-\|_2 + \Sigma \|(F_{\pm}G_{\pm}) \circ z\|_2 \, .$$

Each of the four terms on the right is dominated by $(1 + \|T_0\|^2_{4,4})$ when $\|f\|_4 \leqq 1$ and and $\|g\|_4 \leqq 1$, by Schwarz's inequality. The first two terms are controlled by the same quantity, because of the Calderón's equation. Note that since f and g are compactly supported, the a priori conditions $F_+ = O(\frac{1}{|z|})$ and $G'_+ = O(\frac{1}{|z|})$ as $z \to \infty$ are satisfied.

Thus (2) above follows from the Calderón's equation.

Part 2: <u>Treatment of Calderón's Equation.</u>

An important aspect of Calderón's equation is its conformal invariance. Indeed, if Φ is a conformal mapping from π_+ onto Ω_+, then the equation

$$H'_+ = F_+G'_+$$

is equivalent to

$$(H_+ \circ \Phi)' = (F_+ \circ \Phi)(G_+ \circ \Phi)' \; .$$

Let $\omega = |\Phi'|$, so that $\omega \in A_2$, as we have seen. For $f \in L^p(\Gamma, ds)$,

$$\|f\|_{L^p(\Gamma, ds)} = \|f \circ \Phi\|_{L^p(\omega dx)} \; .$$

Thus,

$$\|H_+\|_{L^2(\Gamma, ds)} \leqq C\|F_+\|_{L^4(\Gamma, ds)}\|G_+\|_{L^4(\Gamma, ds)}$$

will follow from

$$\|H_+\|_{L^2(\omega dx)} \leqq C\|F_+\|_{L^4(\omega dx)}\|G_+\|_{L^4(\omega dx)} .$$

Here we abuse our notation by letting $H_+, F_+,$ and G_+ be functions analytic in π_+ for the second inequality, but we still assume that $H_+, F_+,$ and G_+ satisfy the Calderón's equation.

Let $h_+, f_+,$ and g_+ be the restrictions of $H_+, F_+,$ and G_+ to R. Then their Fourier transforms are supported in $[0,\infty)$, and we have

$$\hat{h}_+(\xi) = \int_0^\xi \hat{f}_+(\xi - \eta)\left(\frac{\eta}{\xi}\right)\hat{g}_+(\eta)d\eta ,$$

by standard Fourier transform formulas.

Let $\varphi \in \mathcal{S}(R)$ satisfy $\varphi(x) = e^{-x}$ for $x \geqq 0$. Then

$$e^{-x} = \int_{-\infty}^{\infty} e^{2\pi i\alpha x}\hat{\varphi}(\alpha)d\alpha ,$$

which, for $0 < \eta \leqq \xi$, implies that

$$\left(\frac{\eta}{\xi}\right) = \int_{-\infty}^{\infty}\left(\frac{\xi}{\eta}\right)^{2\pi i\alpha}\hat{\varphi}(\alpha)d\alpha ,$$

and, hence

$$\hat{h}_+(\xi) = \int_{-\infty}^{\infty}\int_0^\xi \left(\frac{\xi}{\eta}\right)^{2\pi i\alpha}\hat{\varphi}(\alpha)\hat{f}_+(\xi - \eta)\hat{g}_+(\eta)d\eta d\alpha .$$

The multiplier having symbol $(\xi)^{2\pi i\alpha}$ is, according to the theorem on p. 73, a C-Z convolution operator with C-Z norm dominated by $C(1 + |\alpha|)^3$. Let us denote this multiplier by M_α, so that

$$\hat{h}_+ = \int_{-\infty}^{\infty}(\xi)^{2i\pi\alpha}\widehat{\varphi(\alpha)}\,\widehat{(M_{-\alpha}g_+)} * \hat{f}_+ d\alpha .$$

Hence,

$$h_+ = \int_{-\infty}^{\infty}\hat{\varphi}(\alpha)M_\alpha[f_+(M_{-\alpha}g_+)]d\alpha .$$

Since $\omega \in A_2 \subseteq A_4$, each M_α is bounded on $L^2(\omega dx)$ and $L^4(\omega dx)$ with appropriate norm control, and thus

$$\|h_+\|_{L^2(\omega dx)} \leq C \int_{-\infty}^{\infty} |\hat{\varphi}(\alpha)| (1+|\alpha|)^3 \| [f_+ (M_{-\alpha} g_+)] \|_{L^2(\omega dx)} d\alpha$$

$$\leq C \int_{-\infty}^{\infty} |\hat{\varphi}(\alpha)| (1+|\alpha|)^3 \|f_+\|_{L^4(\omega dx)} \|M_{-\alpha} g\|_{L^4(\omega dx)} d\alpha$$

$$\leq C \int_{-\infty}^{\infty} |\hat{\varphi}(\alpha)| (1+|\alpha|)^6 \|f_+\|_{L^4(\omega dx)} \|g_+\|_{L^4(\omega dx)} d\alpha$$

$$\leq C \|f_+\|_{L^4(\omega dx)} \|g_+\|_{L^4(\omega dx)} .$$

The constant C depends only on the A_2-constant of ω.

The above calculations are purely formal, since, to avoid breaking the line of proof, we did not justify the convergence of any of the above integrals. A natural way of making the above computations rigorous is to consider the functions $H_+(\cdot + iy)$, $F_+(\cdot + iy)$, and $G_+(\cdot + iy)$ for $y > 0$, and then let y tend to 0. For then the Fourier transform of the corresponding function h_y is given by $\hat{h}_y = e^{-2\pi|\xi|}\hat{h}_+$, and similarly for f_y and g_y, and these functions can be appropriately controlled. The details are easy and left to the reader.

This finishes the proof of Calderón's theorem. In the next chapter, we shall see how to remove the restriction on δ. For the moment let us consider the following application of Calderón's theorem (which holds for general φ, since, as just mentioned, we shall remove the restriction on δ in the next chapter).

Proposition: Let Γ be the graph of a Lipschitz function φ, and let Ω_+ be the connected component of $\mathbb{C}\setminus\Gamma$ which lies above Γ. For $f \in L^p(\Gamma)$, $1 < p < \infty$, define $F : \Omega_+ \to \mathbb{C}$ by

$$F(z) = \frac{1}{2\pi i} \int_{\mathbb{R}} \frac{f(z(s))}{z(s) - z} z'(s) ds .$$

Then the non-tangential limit of F to the boundary exists almost everywhere, and it is equal to

$$\frac{1}{2} f(z(t)) + \text{P.V.} \frac{1}{2\pi i} \int \frac{f(z(s))}{z(s) - z(t)} z'(s) ds .$$

Here, the non-tangential convergence is defined to mean

$$\lim_{\substack{z \to z(t) \\ (\mathrm{Im}\,z - \mathrm{Im}\,z(t)) > \beta(\mathrm{Re}\,z - \mathrm{Re}\,z(t))}} \quad ,$$

for some fixed β, independent of t, which satisfies $\beta > \|\varphi'\|_\infty$.

As usual, the proof is divided into two steps. The first is to prove the boundedness of the maximal operator associated to this type of convergence, which turns out to be a consequence of the boundedness of C_*, where C is the Cauchy operator associated to Γ. In the second step, one proves the result for a dense class of functions, namely $C_c^\infty(\mathbb{R})$. The proof is left as an exercise.

CHAPTER 8

Some Techniques to Generate New Calderón-Zygmund Operators.

As we have mentioned before, it is usually very difficult to show that a given standard kernel defines a CZO , when it is not a convolution kernel. The previous chapter gives an example of how difficult such a problem can be. Thus it is good to have available as many techniques as possible for generating new operators from old ones, with the hope that these new ones include precisely those that one is trying to handle.

The techniques that we shall present here are all somewhat related to the Cauchy kernel. In particular, it was a conjecture of A.P. Calderón's that the restriction $\|\varphi'\|_\infty < \delta$ (in the theorem of the preceding chapter) could be replaced by $\|\varphi'\|_\infty < \infty$. Indeed, this was proved by Coifman, MacIntosh, and Meyer using a method completely different from that of Calderón. A bit later, Guy David discovered a real variable technique which also permits the restriction $\|\varphi'\|_\infty < \delta$ to be removed, and it uses a method of perturbation. His theorem is quite striking:

Theorem: Let K be a standard kernel on $R^n \times R^n \setminus \Delta$ with corresponding maximal operator T_* . Suppose that there exists $\nu \in (0,1]$ and $C_0 > 0$ such that for all cubes $Q \subseteq R^n$ there exists a measurable subset E of Q such that $|E| > \nu|Q|$; moreover, there exists a standard kernel K_Q satisfying $C(K_Q) \leqq C_0$, $\|(T_Q)_*\|_{2,2} \leqq C_0$, and

$$K_Q\Big|_{E\times E\setminus\Delta} = K\Big|_{E\times E\setminus\Delta} ,$$

where $(T_Q)_*$ is the maximal operator associated with K_Q . Then T_* is bounded on $L^2(R^n, dx)$.

The proof consists of establishing the following good λ's inequality: There exists $\alpha \in (0,1)$ such that for each $\epsilon > 0$, there is a $\gamma > 0$ so that

$$|\{x \in R^n : T_*f(x) > (1+\epsilon)\lambda,\ f^*(x) < \gamma\lambda\}| \leqq \alpha|\{x \in R^n : T_*f(x) > \lambda\}| ,$$

for all $\lambda > 0$. There is no problem in using such a good λ's inequality, since we only need an estimate on $\|T_*f\|_2$ for f bounded and compactly supported, and, as we have seen earlier, such f satisfy $\inf(1, |T_*f|) \in L^2(R^n, dx)$.

For $\lambda > 0$ fixed, the set $\Omega_\lambda = \{x \in R^n : T_* f(x) > \lambda\}$ is open, and it is enough to prove that the good λ's inequality holds when restricted to each maximal dyadic cube contained in Ω_λ. Let Q be any such cube. Although we shall fix γ later, we may assume that $\{x : f^*(x) < \gamma\lambda\} \cap Q$ is not empty (since, otherwise, the desired inequality is trivial).

We are going to show that there exists a subset F of E such that $\frac{|F|}{|Q|} \geq \frac{\nu}{2}$ and $(T_* f) \leq (1+\epsilon)\lambda$ on F. This will imply the good λ's inequality with $\alpha = (1 - \frac{\nu}{2})$.

Let N be large (to be chosen later), let $Q' = (1 - \frac{\nu}{N})Q$, and let $E_1 = E \cap Q'$. (The role of Q' will become clear later in the proof.)

<u>Claim</u>: <u>It is enough to prove that there exists a function</u> $R(x)$ <u>and a constant</u> C_ν <u>such that for</u> $x \in E_1$,

$$T_* f(x) \leq \lambda + (T_Q)_*(f\chi_Q)(x) + C_\nu \gamma\lambda + R(x)$$

<u>where</u> $\int_E R(x)dx \leq C_\nu \gamma\lambda |Q|$.

Indeed, let

$$F = F_1 \cap \{x \in E : (T_Q)_*(f\chi_Q)(x) \leq \frac{\epsilon\lambda}{4}\} \cap \{x \in E : R(x) \leq \frac{\epsilon\lambda}{4}\} .$$

Then

$$|F| \geq |E_1| - |\{x \in R^n : (T_Q)_*(f\chi_Q)(x) > \frac{\epsilon\lambda}{4}\}| - |\{x \in E : R(x) > \frac{\epsilon\lambda}{4}\}| .$$

For N large enough, $|E_1| > \frac{9}{10}|E|$, and

$$|\{x \in E : R(x) > \frac{\epsilon\lambda}{4}\}| \leq \frac{\nu}{10}|Q|$$

if γ is chosen small enough. To estimate

$$|\{(T_Q)_*(f\chi_Q)(x) > \frac{\epsilon\lambda}{4}\}| ,$$

we use the weak type (1,1) inequality on $(T_Q)_*$, and the fact that $m_Q|f| \leq f^*(x) < \gamma\lambda$, to obtain

$$|\{x \in R^n : (T_Q)_*(f\chi_Q)(x) > \frac{\epsilon\lambda}{4}\}| \leq \frac{4c}{\epsilon\lambda}\|f\chi_Q\|_1$$

$$\leq 4c\,|Q|\frac{\gamma}{\epsilon} .$$

Choosing γ small enough, we obtain an estimate of $\frac{\nu}{10}|Q|$ for this set, so that $|F| > \frac{7}{10}\nu|Q|$, as desired.

To prove (1), it suffices to prove the following inequalities:

There exists $C_\nu > 0$ such that

(2) $$T_*(f\chi_{Q^c})(x) \leqq \lambda + C_\nu \gamma\lambda$$

for $x \in Q'$.

There exists $C'_\nu > 0$ such that

(3) $$T_*(f\chi_Q)(x) \leqq (T_Q)_*(f\chi_Q)(x) + R(x)$$

for $x \in E_1$ and with R as above.

To prove (2), we use the fact that Q is maximal in Ω_λ, so that there exists $a \in \tilde{Q}$ such that $T_*f(a) \leqq \lambda$. (Recall that $\tilde{Q}$ is the "double" of Q with respect to the dyadic mesh; see Section II, Chapter 1.) On the other hand, for each cube $\hat{Q}$ containing Q we have $m_{\hat{Q}}|f| \leqq \gamma\lambda$. With these remarks it is easy to use the standard estimates on K to show that there exists $C_\nu > 0$ such that

$$T_*(f\chi_{(\tilde{Q})^c})(a) \leqq T_*f(a) + C_\nu\gamma\lambda \quad ,$$

$$T_*(f\chi_{Q^c})(x) \leqq T_*(f\chi_{(\tilde{Q})^c})(x) + C_\nu\gamma\lambda \quad \text{for} \quad x \in Q' \quad ,$$

and

$$T_*(f\chi_{(\tilde{Q})^c})(x) \leqq T_*(f\chi_{(\tilde{Q})^c})(a) + C_\nu\gamma\lambda \quad .$$

(The second inequality is restricted to Q' so that x is significantly distant from the support of $f\chi_{Q^c}$.) As is readily seen, these three inequalities imply (2).

To prove (3) we begin by observing that for $x \in E_1$ and $\eta > 0$,

$$|T_\eta(f\chi_Q)(x) - (T_Q)_\eta(f\chi_Q)(x)| \leqq \int_Q |K(x,y) - K_Q(x,y)|\,|f(y)|\,dy \ .$$

To estimate this, we use the following.

<u>Lemma</u>: For $x \in E$ <u>and</u> $y \in Q$,

$$|K(x,y) - K_Q(x,y)| \leqq \frac{Cd(y,E)}{|x-y|^{n+1}} \quad .$$

For, if $|x-y| \leq 10\, d(y,E)$, then

$$|K(x,y) - K_Q(x,y)| \leq \frac{2C}{|x-y|^n} \leq \frac{cd(y,E)}{|x-y|^{n+1}}$$

If $|x-y| \geq 10\, d(y,E)$, then let $z \in E$ be such that $d(y,E) = d(y,z)$. (We can assume that E is closed, so that z exists.) Because z and x lie in E , $|K(x,z) - K_Q(x,z)| = 0$, and, applying the standard estimates on K and K_Q , we obtain

$$|K(x,y) - K_Q(x,y)| \leq \frac{Cd(y,z)}{|x-y|^{n+1}} = \frac{Cd(y,E)}{|x-y|^{n+1}} .$$

In view of the lemma, we can define R by

$$R(x) = \int_Q \frac{Cd(y,E)}{|x-y|^{n+1}} |f(y)|\, dy .$$

Hence,

$$\int_E R(x)dx \leq C \int_Q [\int_E \frac{d(y,E)}{|x-y|^{n+1}} dx] |f(y)|\, dy$$

$$\leq C \int_Q [\int_{R^n \setminus B(y,d(y,E))} \frac{d(y,E)}{|x-y|^{n+1}} dx] |f(y)|\, dy$$

$$\leq C \int_Q |f(y)|\, dy \leq C|Q|\gamma\lambda .$$

This proves (3), and the proof of the theorem is concluded.

This theorem, combined with the following lemma, allows us to remove the restriction $\|\varphi'\|_\infty < \delta$ from Calderón's theorem.

__Lemma__: __Let__ Γ __be the graph of a Lipschitz function__ φ __satisfying__ $\|\varphi'\|_\infty \leq M$. __Let__ $z(\cdot)$ __be an arclength parametrerization of this curve, and let__ I __be an interval on__ R . __Then there exists a curve__ Γ_1 __which is isometric to the graph of a Lipschitz function__ φ_1 __satisfying__ $\|\varphi_1'\|_\infty \leq \frac{19}{20}M$, __and there exists a parameterization__ z_1 __of__ Γ_1 __which satisfies__:

(i) $z_1 = z$ __on a measurable subset__ E __of__ I , $\frac{|E|}{|I|} \geq \frac{1}{3}$;

(ii) $C \leq |z_1'(s)| \leq 1$; __where__ $C > 0$ __is independent of__ M .

From this lemma and the preceding theorem it follows that if $\|\varphi'\|_\infty < \delta$ implies that $\frac{1}{z(x)-z(y)}$ defines an operator on L^2 of norm less than M_δ, then $\|\varphi'\|_\infty \leqq \frac{20}{19}\delta$ implies that the kernel $\frac{1}{z(x)-z(y)}$ defines a bounded operator of norm less than CM_δ, where C is an absolute constant. Thus the restriction $\|\varphi'\|_\infty < \delta$ can be removed, and we get an estimate of the form $C(1+\|\varphi'\|_\infty)^N$ for the CZ norm of the kernel $\frac{1}{z(x)-z(y)}$.

Let us now sketch the proof of this geometric lemma. Details and justifications are elementary and left to the reader.

I being fixed, let J be the interval in R which satisfies $\{x+i\varphi(x) : x \in J\} = z(I)$, and let

$$U_0 = \{x \in J : \sup_{x<y\in J} m_{[x,y]}\varphi' > \frac{9}{10}M\}$$

and

$$U_1 = \{x \in J : \inf_{x<y\in J} m_{[x,y]}\varphi' < -\frac{9}{10}M\} .$$

Then U_0 and U_1 are both open subsets of J made up of intervals (x,y) such that $m_{[x,y]}\varphi \geqq \frac{9}{10}M$ (respectively $m_{[x,y]}\varphi' \leqq -\frac{9}{10}M$). (This follows from the so called "rising sun lemma"; see [Z].) Hence, for each connected component of U_0, say J_K, $\varphi' > 0$ on at least $\frac{9}{10}$ of J_K, and $\varphi' \leqq 0$ on at most $\frac{1}{10}$ of J_K.

Let I_K be the subinterval of I corresponding to J_K, so that $|I_K| \geqq |J_K|\frac{9}{10}\sqrt{1+M^2}$. The subset of I_K corresponding to the subset of J_K of the x's for which $\varphi'(x) \leqq 0$ has a measure less than $\frac{1}{10}|J_K|\sqrt{1+M^2}$, and hence, less than $\frac{1}{9}|I_K|$. From this, one can see that if V_0 and V_1 are the subsets of I corresponding to U_0 and U_1, then

$$\frac{|V_0|}{|I|} \leqq \frac{2}{3} \quad \text{or} \quad \frac{|V_1|}{|I|} \leqq \frac{2}{3} .$$

Suppose, for instance, that $\frac{|V_1|}{|I|} \leqq \frac{2}{3}$. And let z_1 be such that:

(i) $z_1 = z$ on $I \setminus V_1$,

(ii) $z_1' = 1$ outside of I,

(iii) z' is constant on the connected components of V_1.

To see that z_1 has the desired properties, we observe that

$$-\arctan\left(\frac{9}{10}M\right) \leqq \operatorname{Arg} z_1' \leqq \arctan(M) .$$

Hence the curve Γ_1 described by z_1 is isometric to the graph of a function φ_1 such that

$$\|\varphi_1'\|_\infty \leqq \tan\left(\frac{1}{2}\arctan M + \frac{1}{2}\arctan\frac{9}{10}M\right)$$

$$\leqq \frac{19}{20}M .$$

(Γ_1 will be the image of the graph of φ_1 under a rigid motion of the plane.) On the other hand, $|z_1'| > \frac{9}{10}$. This proves the lemma.

Besides the techniques used above, which make full use of the Calderón-Zygmund machinery, there are simpler ones which nevertheless have interesting applications. The following, due to Coifman, are examples of this. One such technique and the proposition on p. 63 (see Chapter 4, Section V) will be used to establish:

<u>Proposition 1</u>: <u>Let</u> $A : \mathbb{R} \to \mathbb{R}$ <u>satisfy</u> $\|A'\|_\infty < \infty$. <u>Then the kernel</u>

$$K(x,y) = \left[\exp\left(i\,\frac{A(x) - A(y)}{x - y}\right)\right]\frac{1}{x - y}$$

<u>is a</u> CZK <u>with norm dominated by</u> $C(1 + \|A'\|_\infty)^N$, <u>for some</u> $C, N > 0$.

<u>Proof</u>: Let Γ be the following contour in $\mathbb{C}$.

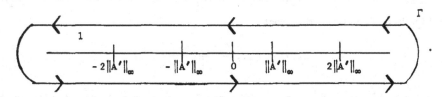

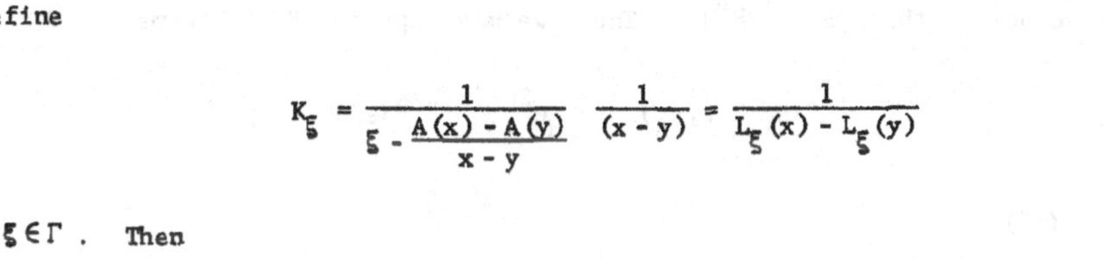

We define

$$K_\xi = \frac{1}{\xi - \frac{A(x) - A(y)}{x - y}}\ \frac{1}{(x - y)} = \frac{1}{L_\xi(x) - L_\xi(y)}$$

for $\xi \in \Gamma$. Then

$$K(x,y) = \int_\Gamma K_\xi(x,y)e^{i\xi}\,\frac{d\xi}{2\pi i} .$$

Hence, it is enough to show that for each $\xi \in \Gamma$, $K_\xi(\cdot,\cdot)$ is a CZK, and that

$$\int_\Gamma \|K_\xi\|_{CZ} d|\xi| \leq C(1+\|A'\|_\infty)^N .$$

Let us consider first the case where $\xi = u \pm i$, $u \in [-2\|A'\|_\infty, 2\|A'\|_\infty]$. Then $L_\xi(x) = (u \pm i)x - A(x) = \pm i\,[x \mp i(ux - A(x))]$. Since $|u| + \|A'\|_\infty \leq 3\|A'\|_\infty$, K_ξ is a CZK with norm less than $C(1+\|A'\|_\infty)^N$. (See the remarks following the statement of the preceding lemma.)

If $\xi = u+iv$ with $|v| < 1$ and $|u| > 2\|A'\|_\infty$, then

$$C_\xi(x) = \xi\left(x - \frac{A(x)}{\xi}\right) = \xi\left[\operatorname{Re}\left(x - \frac{A(x)}{\xi}\right) - i \operatorname{Im}\frac{A(x)}{\xi}\right] .$$

Applying the change of variables $x' = \operatorname{Re}\left(x - \frac{A(x)}{\xi}\right)$, and using the fact that $\frac{\|A'\|_\infty}{|\xi|} < \frac{1}{2}$, one can see readily that $\|K_\xi\|_{CZ} \leq \frac{C}{|\xi|}$. The desired estimate on $\int_\Gamma \|K_\xi\|_{CZ} d|\xi|$ now follows immediately.

Another example of such a technique is used to establish:

<u>Proposition 2</u>: *Let* $F : \mathbb{R}^n \to \mathbb{C}$ *be* C^∞, *and let* $A_1,\ldots,A_n$ *be Lipschitz functions taking* $\mathbb{R}$ *into itself. Then*

$$\frac{1}{x-y} F\left(\frac{A_1(x) - A_1(y)}{x-y}, \ldots, \frac{A_n(x) - A_n(y)}{x-y}\right)$$

is a CZK .

<u>Proof</u>: Clearly this kernel satisfies the standard estimates. Because $A_1,\ldots,A_n$ are Lipschitz functions, the function

$$(x,y) \to \left(\ldots, \frac{A_i(x) - A_i(y)}{x-y}, \ldots\right)$$

which takes $\mathbb{R}^2$ into $\mathbb{R}^n$ has bounded range. Hence we may change F without changing the kernel so that $F \in C_c^\infty(\mathbb{R}^n)$. Thus, we may express F in terms of $\hat{F}$:

$$F(x_1,\ldots,x_n) = \int_{\mathbb{R}^n} e^{2\pi i \langle x|\xi\rangle} \hat{F}(\xi)\,d\xi ,$$

and $\hat{F} \in \mathcal{S}(\mathbb{R}^n)$.

From Proposition 1 we know that the kernel

$$\frac{1}{(x-y)} \exp\left(2\pi i \sum_{i=1}^{n} \xi_i \frac{A_i(x) - A_i(y)}{x-y}\right)$$

is a CZK with norm dominated by

$$C[1+|\xi|(\sum_{i=1}^{n}\|A'\|_\infty)]^N .$$

Since the integral

$$\int_{R^n}[1+|\xi|(\sum_{i=1}^{n}\|A'\|_\infty)]^N \hat{F}(\xi)d\xi$$

converges, the proposition follows by another application of the proposition in Chapter 4, Section V, p. 63 .

A corollary of Proposition 2 is the following:

<u>Proposition 3</u>: <u>Let</u> Γ <u>be a Jordan curve, and let</u> z <u>be an arclength parameterization of</u> Γ . <u>If there exists</u> $C(\Gamma)>0$ <u>such that</u>

$$|z(x)-z(y)| \geq C(\Gamma)|x-y| ,$$

then $\frac{1}{z(x)-z(y)}$ is a CZK .

The curves which satisfy the above property are called <u>chord-arc curves</u>. This proposition is obtained by setting $A_1 = \text{Re}\, z$, $A_2 = \text{Im}\, z$, and by choosing $F \in C^\infty(R^2)$ such that $F(x,y) = \frac{1}{x+iy}$ for $|x+iy| > C(\Gamma)$.

Note that the chord-arc curves are precisely those for which the kernel $\frac{1}{z(x)-z(y)}$ is a standard kernel. It can be shown, however, that the Cauchy kernel is bounded on $L^p(\Gamma, ds)$, $1<p<\infty$, for more general curves. See [D].

We conclude this short review of techniques which are now part of the Calderón-Zygmund machinery with a specific problem: the L^2 boundedness of multipliers on Lipschitz curves.

In [CM3], Coifman and Meyer study the notion of the Fourier transform on a Lipschitz curve Γ , which is defined by

$$\hat{f}(t) = \int_\Gamma e^{-2\pi i t z} f(z)dz$$

for suitable functions f . Thus, if f is defined on Γ, $\hat{f}$ is defined on R .

Let us now assume that Γ is the graph of a function $\varphi \in C_c^\infty(R)$, that f is holomorphic on a strip containing Γ and that f is uniformly in L^2 on each of the horizontal lines which make up this strip. It is not hard to show that the restriction to Γ of such f's are dense in $L^2(\Gamma)$; see [CM3].

For such functions f , we can prove the inversion formula

$$f(z) = \int_{\mathbb{R}} e^{2\pi i t z} \hat{f}(t)dt .$$

Indeed, since both sides are holomorphic in z , it is enough to prove this equality when z lies on the real axis. But in this case, a change of contour in the formula

$$\hat{f}(t) = \int_{\Gamma} e^{-2\pi i t z} f(z)dz$$

reduces us to the classical case applied to the restriction of f to the real axis.

Now that we have a Fourier inversion formula, it is natural to consider the corresponding multipliers defined by

$$Tf(z) = \int_{\mathbb{R}} e^{2\pi i t z} m(t)\hat{f}(t)dt$$

for $z \in \Gamma$ (of course, this definition is purely formal). Tf(z) can be thought of as the holomorphic extension to Γ of $G : \mathbb{R} \to \mathbb{C}$ defined by

$$G(x) = \int_{\mathbb{R}} e^{2\pi i t x} m(t)\hat{f}(t)dt \quad .$$

For example, suppose that $m(t) = \chi_{\mathbb{R}_+}(t)$. Then,

$$G(x) = \frac{1}{2\pi i} \int_{\mathbb{R}} \frac{1}{x - y} f(y)dy + \frac{1}{2} f(x) \quad ,$$

so that the holomorphic extension of G is

$$G(z) = \frac{1}{2\pi i} \int_{\mathbb{R}} \frac{1}{x - y'} f(y')dy' \quad z \notin \mathbb{R} \ ;$$

hence,

$$Tf(z) = \frac{1}{2\pi i} \int_{\Gamma} \frac{1}{z - z'} f(z') + \frac{1}{2} f(z) ,$$

for $z \in \Gamma$. Thus Tf appears as a convolution operator along Γ , of f and of the holomorphic extension of $\frac{1}{2\pi i z} + \frac{1}{2}\delta$ (where δ is the point mass concentrated at the origin) to the domain $\Gamma - \Gamma$. Note that $\Gamma - \Gamma$ is contained in the sector $\{z : |\operatorname{Im} z| \leq \|\varphi'\|_\infty |\operatorname{Re} z|\}$.

We shall adopt this point of view, since it gives an explicit description of the kernel of T . Let us now state precisely the problem we are going to solve.

Problem: Let m be the symbol of a convolution CZO with kernel K. What are the conditions on m that imply the following

(i) K extends holomorphically, to a sector of the form $\{z : |\mathrm{Im}\, z| < M|\mathrm{Re}\, z|\}$, and

(ii) for each function φ satisfying $\|\varphi'\|_\infty < M$, $K((x+i\varphi(x)) - (y+i\varphi(y)))$ is a CZK whose norm is dominated by a constant that depends only on M ?

This problem is studied in detail in [JO]. Simple arguments show that a necessary condition is that m admits a bounded holomorphic extension to sector of the same type. What we shall see now is how to use the boundedness of the Cauchy kernel to show that this condition is also sufficient.

Let us make some simplifying assumptions, which can be removed using techniques which should now be familiar to the reader.

(1) m vanishes for $x > 0$. (The part of m on $x < 0$ can be treated analogously.)

(2) m satisfies an inequality of the form $|m(z)| \leq Ce^{-\varepsilon|z|^a}$, for some $a > 1$.

(3) $\varphi \in C_c^\infty(\mathbb{R})$.

Of course, all of our estimates will be independent of these three assumptions.

Suppose that m has a bounded holomorphic extension (which we also denote by m) to $\{z : |\mathrm{Im}\, z| < (\tan\alpha)\mathrm{Re}\, z\}$, for some $\alpha \in (0, \frac{\pi}{2})$, and let $\|m\|_\infty$ be the L^∞ norm of this extension.

Assumption (2) implies that the kernel K is given by

$$K(x,y) = \int_0^\infty \exp(2\pi i t(x-y))m(t)dt .$$

Let

$$S(z) = \int_0^\infty \exp(2\pi i t z)m(t)dt ,$$

so that $K(x,y) = S(x-y)$. Since, in assumption (2), $a > 1$, S is an entire function. Furthermore a change of contour shows that

$$S(z) = \int_0^\infty \exp(2\pi i e^{i\beta} t z)m(te^{i\beta})e^{i\beta}dt$$

for every $\beta \in (-\alpha, \alpha)$.

Let $\gamma \in (-\alpha, \alpha)$. Then there exists $C_\gamma > 0$ and $\beta \in (0,\alpha)$ such that $\mathrm{Im}\, z \geq (\tan\gamma)|\mathrm{Re}\, z|$ implies $\mathrm{Im}(e^{i\beta}z) \geq C_\gamma|z|$ and thus,

$$|S(z)| \leq \int_0^\infty \exp(-C_\gamma|z|x)\|m\|_\infty dx \leq \frac{C_\gamma}{|z|}\|m\|_\infty$$

for such z. Hence, if $\|\varphi'\|_\infty < \tan\gamma$, then $K(x+i\varphi(x), y+i\varphi(y))$ satisfies the first standard estimate. For the second estimate, we note that the Cauchy formula implies that $|S'(z)| \leqq \frac{C_\gamma}{|z|^2}\|m\|_\infty$, uniformly on each such γ-sector. Hence S is the desired extension of K. Similarly, $|S''(z)| \leqq \frac{C_\gamma}{|z|^3}\|m\|_\infty$, as can be seen by the same argument.

To prove the L^2 boundedness of the operator associated to this kernel, we need to introduce certain operators on Lipschitz domains, the first of which is the analogue of the Lusin area integral.

Suppose that $f \in L^2(\Gamma)$ is the boundary value function of its holomorphic extension F to Ω_+. Then we define S_1 and S_2 on Γ by

$$S_1 f(z_0) = \Big[\iint_{y_2 > A|y_1|} |F'(z_0 + y_1 + iy_2)|^2 dy_1 dy_2\Big]^{\frac{1}{2}}$$

and

$$S_2 f(z_0) = \Big[\int_0^\infty y^3 |F''(z_0 + iy)|^2 dy\Big]^{\frac{1}{2}} .$$

Carlos E. Kenig proved in [K] a general version of the following result.

<u>Theorem</u>: $\|f\|_{L^2(\Gamma)} \approx \|S_1 f\|_{L^2(\Gamma)} \approx \|S_2 f\|_{L^2(\Gamma)}$.

The constants involved depend only on the Lipschitz constant of the curve Γ. A proof of this theorem is given in appendix.

Now, we want to prove that the function G defined on Γ by

$$G(z_0) = \int_\Gamma S(z_0 - z') f(z') dz'$$

satisfies

$$\|G\|_{L^2(\Gamma)} \leqq c\|f\|_{L^2(\Gamma)} .$$

In order to avoid convergence problems, we assume that $f \in C_c^\infty(\Gamma)$. Using Calderón's theorem (see the preceding chapter), we decompose f in the usual way:

$$f = F_+\Big|_\Gamma + F_-\Big|_\Gamma ,$$

and define G_- by

$$G_-(z_0) = \int_\Gamma S(z_0 - z') F_-(z') dz' .$$

By Cauchy's theorem, we may translate Γ downwards without changing the value of the integral. This integral must therefore be 0, since, for $z' \in \Omega_-$ near ∞, $|S(z_0 - z')| \leqq \frac{C}{|z'|}$ and $|F_-(z')| \leqq \frac{C}{|z'|}$. Thus, $G_- \equiv 0$ on Γ, and we are reduced to showing that $\|G_+\|_2 \leqq C\|f\|_2$, where

$$G_+(z_0) = \int_\Gamma S(z_0 - z')F_+(z')dz' .$$

Because of Kenig's theorem and Calderón's theorem, it is enough to show that

$$|S_2(G_+|_\Gamma)(z_0)| \leqq C|S_1(F_+|_\Gamma)(z_0)| .$$

To estimate $G''_+(z_0 + it)$, we make the following change of contour.

Let Λ be the fixed curve defined by $\operatorname{Im} z = 2M|\operatorname{Re} z|$. Then

$$G_+(z_0 + it) = \int_{\Lambda + z_0 + \frac{it}{2}} S(z_0 + it - z')F_+(z')dz' .$$

The change of contour can be justified as before. Because of the estimates on S' and S'', we can differentiate twice under the integral sign. Thus we obtain

$$G''_+(z_0 + it) = \int_{\Lambda + z_0 + \frac{it}{2}} S''(z_0 + it + z')F_+(z')dz' .$$

(Observe that, in differentiating, we do not have to take into account the dependence on z_0 of the contour, since the integrand is holomorphic.)

If we now integrate by parts, the boundary terms vanish, and

$$G''_+(z_0 + it) = -\int_{\Lambda + z_0 + \frac{it}{2}} S'(z_0 + it - z')F'_+(z')dz' .$$

Hence,

$$|G_+''(z_0+it)| \leq \left[\int_{\Lambda+z_0+\frac{it}{2}} |S'(z_0+it-z')|^2 d|z'|\right]^{\frac{1}{2}} \left[\int_{\Lambda+z_0+\frac{it}{2}} |F_+'(z')|^2 d|z'|\right]^{\frac{1}{2}}$$

Hence,

$$|S_2(G_+|_\Gamma)(z_0)|^2 \leq \int_0^\infty t^3 \left[\int_{\Lambda+z_0+\frac{it}{2}} |S'(z_0+it-z')|^2 d|z'|\right] \left[\int_{\Lambda+z_0+\frac{it}{2}} |F_+'(z')|^2 d|z'|\right] dt$$

Using the estimate $|S'(z)| \leq \frac{C}{|z|^2}$ (which holds uniformly throughout each region $\{z : \operatorname{Im} z \geq (\tan\gamma)|\operatorname{Re} z|\}$, $\gamma<\alpha$), we dominate the first factor by $\frac{C}{t^3}$. The left side is dominated by

$$C\int_0^\infty \int_{\Lambda+z_0+\frac{it}{2}} |F_+'(z')|^2 d|z'|\, dt \ .$$

Up to a multiplicative factor, this is $|S_1(F_+|_\Gamma)(z_0)|^2$, which is what we wanted.

This concludes our presentation of multipliers on a Lipschitz curve.

APPENDIX

A Proof of Kenig's Theorem.

Note first that the proof yields a result less general that Kenig's result as cited in [K] . Indeed we make heavy use of the fact that we are dealing with L^2-norms.

Let us begin with the first equivalence. We have

$$\|S_1 f\|^2_{L^2(\Gamma)} = \iint_{\Omega_+} |f'(z)|^2 \alpha(z) dy_1 dy_2 ,$$

where $z = y_1 + iy_2$,

$$\alpha(z) = \int_\Gamma \chi_{\Lambda_{z_0}}(z) |d z_0| ,$$

and Λ_{z_0} is the appropriate cone with vector at z_0 . (See the figure.)

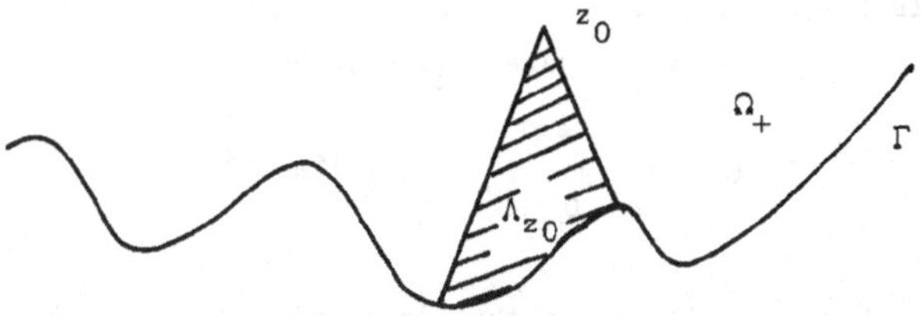

It is clear that $\alpha(z) \approx t(z)$, where $t(z) > 0$ is defined by $z - it(z) \in \Gamma$. Hence,

$$\|S_1 f\|^2_{L^2(\Gamma)} \approx \iint_{\Omega_+} |f'(z)|^2 t(z) dy_1 dy_2 .$$

As in the last chapter, let Φ be a nice conformal mapping from π_+ to Ω_+ . Then,

$$\|S_1 f\|^2_{L^2(\Gamma)} \approx \iint_{\pi_+} |f' \circ \Phi|^2 t(\Phi(z)) |\Phi'(z)|^2 dxdy ,$$

and by Lemma 2 on p.118 in the preceding chapter, we obtain

$$\|S_1 f\|^2_{L^2(\Gamma)} \approx \iint_{\pi_+} |f' \circ \Phi|^2 [\int_{x-y}^{x+y} |\Phi'(t)| dt] |\Phi'(z)|^2 dxdy .$$

Exchanging the order of integration, we see that the left side equals

$$\|S_1(f\circ\Phi)\|_{L^2(|\Phi'|dx)} \quad ,$$

where here S_1 is the Lusin function corresponding to the upper half plane, with $A=1$, (Recall that, in the definition of S_1, A showed up as the apeture of an appropriate cone.) This function is essentially the same as the S-function in Section II of Chapter 6, and hence the results of that chapter apply. In particular, since $|\Phi'|\in A_2$ (by Lemma 1 of the last chapter), we may apply the proposition in Section I of Chapter 6 to conclude that

$$\|S_1(f\circ\Phi)\|_{L^2(|\Phi'|dx)} \approx \|f\circ\Phi\|_{L^2(|\Phi'|dx)}$$

$$\approx \|f\|_{L^2(\Gamma,ds)} \quad ,$$

which implies the first equivalence.

For the second equivalence, we replace $S_1f(z_0)$ by

$$S_1'f(z_0) = \left[\int_0^\infty y|f'(z_0+iy)|^2dy\right]^{\frac{1}{2}} .$$

Because $\|S_1f\|_{L^2(\Gamma)} \approx \|S_1'f\|_{L^2(\Gamma)}$, as in the classical case, it is enough to show that

$$\|S_1'f\|_{L^2(\Gamma)} \approx \|S_2f\|_{L^2(\Gamma)} .$$

As in the case where Γ is the real line, it is not hard to show the pointwise inequality $|S_1'f(z_0)| \leqq C|S_2f(z_0)|$. Indeed, it is easy to show that for each z_0, $\lim_{y\to\infty} f''(z_0+iy)=0$, hence

$$f'(z_0+iy) = -\int_y^\infty f''(z_0+i\eta)d\eta \quad .$$

Therefore,

$$|f'(z_0+iy)|^2 \leqq \left[\int_y^\infty \eta^2|f''(z_0+i\eta|^2d\eta\right]\left[\int_y^\infty \frac{d\eta}{\eta^2}\right]$$

$$\leqq \frac{1}{y}\left[\int_y^\infty \eta^2|f''(z_0+i\eta)|^2d\eta\right] \quad ,$$

Minkowski's integral inequality implies now:

$$\int_0^\infty y|f'(z_0+iy)|^2dy \leqq \int_0^\infty \eta^2|f''(z_0+i\eta)|^2 [\int_0^\eta dy]d\eta .$$

Hence $|S_1'f(z_0)|^2 \leqq |S_2f(z_0)|^2$, which is what we desired.

The converse inequality is not quite as easy. (Note that it is the one that we used above.) To prove it, we shall make use of the following geometric observation, valid for all Lipschitz curves Γ : there exists $a>0$ so that for all $z \in \Omega_+$,

$$\{\zeta : \frac{y(z)}{2a} < |\zeta - z| < \frac{y(z)}{a}\}$$

$$\subseteqq \{\zeta : \frac{y(z)}{2} < y(\zeta) < \frac{3y(z)}{2}\} ,$$

where $y(z)>0$ is defined by $z - iy(z) \in \Gamma$.

Using the Cauchy formula for derivatives, one can easily show that

$$|f''(z)| \leqq \frac{C}{y(z)^3} \iint_{\frac{y(z)}{2a} \leqq |\zeta - z| \leqq \frac{y(z)}{a}} |f'(\zeta)|d\zeta_1 d\zeta_2 ,$$

where $\zeta = \zeta_1 + i\zeta_2$. Thus,

$$y(z)^3|f''(z)|^2 \leqq C \iint_{\frac{y(z)}{2a} \leqq |\zeta - z| \leqq \frac{y(z)}{a}} \frac{|f'(\zeta)|^2}{y(\zeta)} d\zeta_1 d\zeta_2 .$$

Let $S(\zeta) = |\{z : \frac{y(z)}{2a} < |\zeta - z| < \frac{y(z)}{a}\}|$. Then $S(\zeta) < Cy(\zeta)^2$, by the above observation. Therefore, with $z = z_1 + iz_2$,

$$\|S_2 f\|_2 = \iint_{\Omega_+} y(z)^3|f''(z)|^2 dz_1 dz_2$$

$$\leqq C \iint_{\Omega_+} \frac{S(\zeta)}{y(\zeta)} |f'(\zeta)|^2 d\zeta_1 d\zeta_2$$

$$\leqq C \iint_{\Omega_+} \frac{y(\zeta)^2}{y(\zeta)} |f'(\zeta)|^2 d\zeta_1 d\zeta_2$$

$$\leqq C\|S_1 f\|_2$$

This completes the proof of Kenig's theorem.

REFERENCES

[B-G] D.L. Burkolder and R.F. Gundy. "Extrapolation and interpolation of quasi-linear operators on martingales". Acta Math., Vol. 124 (1970), 249-304.

[C1] A.P. Calderón. "Commutators of singular integral operators". Proc. Nat. Acad. Sci. U.S.A., 53 (1965), 1092-1099.

[C2] ____________. "Cauchy integrals on Lipschitz curves and related operators". Proc. Nat. Acad. Sci. U.S.A., 74 (1977), 1324-1327.

[C3] ____________. "Commutators, singular integral on Lipschitz curves and applications". Proc. of the I.C.M. Helsinki (1978).

[CCFJR] A.P. Calderón, C.P. Calderón, E.B. Fabes, M. Jodeit, N.M. Riviere. "Applications of the Cauchy integral along Lipschitz curves". Bull. Amer. Math. Soc., Vol. 84, No. 2 (1978), 287-290.

[CZ] A.P. Calderón and A. Zygmund. "On the existence of certain singular integrals". Acta Math. 88 (1952), 85-139.

[CF] R.R. Coifman and C. Fefferman. "Weighted norm inequalities for maximal functions and singular integrals". Studia Math. 51 (1974), 241-250.

[CMcIM] R. Coifman, A.McIntosh, Y.Meyer. "L'integrale de Cauchy definit un opérateur borné sur L^2 pour les courbes Lipschitziennes". Ann. of Math. 116 (1982) 361-388.

[CM1] R.R. Coifman and Y. Meyer. "Le theorème de Calderón par les méthodes de variable reelle". C.R. Acad Sci. Paris 289 (1979), 425-428.

[CM2] ________________________. "Au dela des operateurs pseudo-differentiels". Asterisque 57 (1978).

[CM3] ________________________. "Fourier analysis of multilinear convolutions, Calderón's theorem, and analysis on Lipschitz curves". Lectures Notes in Math. No. 779 (1979), 109-122

[CR] R.R. Coifman and R. Rochberg. "Another characterization of BMO". Proc. Amer. Math. Soc., Vol. 79, No. 2 (1980), 249-254.

[CRW] R.R. Coifman, R. Rochberg, G. Weiss. "Factorization theorems for Hardy spaces in several variables". Ann. of Math., 103 (1976), 611-635.

[CW1] R.R. Coifman and G. Weiss. "Extensions of Hardy spaces and their use in analysis". Bull. Amer. Math. Soc., 83 (1977), 569-645.

[CW2] ________________________. Book review of "Littlewood-Paley and multiplier theory" of R.E. Edwards and G.I. Gaudry. Bull. Amer. Math. Soc., Vol. 84 (1978), 242-250.

[CO] A. Cordoba. "Translation invariant operators". Proc. Seminar at El Escorial, edited by M. Guzman and E. Peral (1979), 117-176.

[COF] A. Cordoba and R. Fefferman. "On the equivalence between the boundedness of certain classes of maximal and multiplier operators in Fourier analysis". Proc. Nat. Acad. Sci. U.S.A., Vol. 74 (1977), 423-425.

[D] G. David. Private communication.

[F] C. Fefferman. "Recent progress in classical Fourier analysis". Proc. of the I.C.M. Vancouver (1974).

[FS] C. Fefferman and E.M. Stein. "H^p spaces of several variables". Acta Math. 129 (1972), 137-193.

[G] J. Garcia-Cuerva. "An extrapolation theorem in the theory of A_p-weights". To appear.

[GU] M. de Guzman. "Differentiation of integrals in R^n". Lecture Notes 481 Springer Verlag (1975).

[H] L. Hörmander. "Pseudo-differential operators and hypoelliptic equations". Proc. Symp. Pure Math., 10 (1966), 138-183.

[J] M. Jodeit, Jr. "Applications of A.P. Calderón's Cauchy integral theorem". Proc. Symp. Pure Math., Vol. XXXV, Part 2 (1979), 197-201.

[JN] F. John and L. Nirenberg. "On functions of bounded mean oscillation". Comm. Pure Appl. Math., 18 (1965), 415-426.

[JO] J.L. Journé. "Conjugaison d'operateurs pseudo-differentiels par des diffeomorphismes preservant BMO". Thesis Orsay No. 2955 (1981).

[K] J.P. Kahane. "Some random series of functions". Heath Math. Monographs (1968).

[KE] C.E. Kenig. "Weighted Hardy spaces on Lipschitz domains". Amer. J. Math., 102 (1980), 129-163.

[M] B. Muckenhoupt. "Weighted norm inequalities for the Hardy maximal function". Trans. Amer. Math. Soc., 165 (1972), 207-226.

[MU] N.T. Mushkelishvili. "Singular integral equations". P. Noordhoff, Groningen, 1953.

[NSW] A. Nagel, E.M. Stein, S. Wainger. "Differentiation in lacunary directions". Proc. Nat. Acad. Sci. U.S.A., Vol. 75 (1968), 1060-1062.

[P] J. Plemelj. Monatsh. Math. Phys. 19 (1908), 205-210, 211-245.

[R] J. Rubio de Francia. "Factorization and extrapolation of weights". Bull. Amer. Math. Soc. 7 (1982), 393-396.

[S] E.M. Stein. "Singular integrals and differentiability properties of functions." Princeton University Press (1970).

[SW] E.M. Stein and G. Weiss. "Fourier analysis on Euclidean spaces". Princeton University Press (1971).

[Y] K. Yosida. "Functional analysis". Springer-Verlag, (1965).

[Z] A. Zygmund. "Trigonometric series". 2 Vols. Cambridge (1968).

INDEX

Vol. 845: A. Tannenbaum, Invariance and System Theory: Algebraic and Geometric Aspects. X, 161 pages. 1981.

Vol. 846: Ordinary and Partial Differential Equations, Proceedings. Edited by W. N. Everitt and B. D. Sleeman. XIV, 384 pages. 1981.

Vol. 847: U. Koschorke, Vector Fields and Other Vector Bundle Morphisms - A Singularity Approach. IV, 304 pages. 1981.

Vol. 848: Algebra, Carbondale 1980. Proceedings. Ed. by R. K. Amayo. VI, 298 pages. 1981.

Vol. 849: P. Major, Multiple Wiener-Itô Integrals. VII, 127 pages. 1981.

Vol. 850: Séminaire de Probabilités XV. 1979/80. Avec table générale des exposés de 1966/67 à 1978/79. Edited by J. Azéma and M. Yor. IV, 704 pages. 1981.

Vol. 851: Stochastic Integrals. Proceedings, 1980. Edited by D. Williams. IX, 540 pages. 1981.

Vol. 852: L. Schwartz, Geometry and Probability in Banach Spaces. X, 101 pages. 1981.

Vol. 853: N. Boboc, G. Bucur, A. Cornea, Order and Convexity in Potential Theory: H-Cones. IV, 286 pages. 1981.

Vol. 854: Algebraic K-Theory. Evanston 1980. Proceedings. Edited by E. M. Friedlander and M. R. Stein. V, 517 pages. 1981.

Vol. 855: Semigroups. Proceedings 1978. Edited by H. Jürgensen, M. Petrich and H. J. Weinert. V, 221 pages. 1981.

Vol. 856: R. Lascar, Propagation des Singularités des Solutions d'Equations Pseudo-Différentielles à Caractéristiques de Multiplicités Variables. VIII, 237 pages. 1981.

Vol. 857: M. Miyanishi. Non-complete Algebraic Surfaces. XVIII, 244 pages. 1981.

Vol. 858: E. A. Coddington, H. S. V. de Snoo: Regular Boundary Value Problems Associated with Pairs of Ordinary Differential Expressions. V, 225 pages. 1981.

Vol. 859: Logic Year 1979-80. Proceedings. Edited by M. Lerman, J. Schmerl and R. Soare. VIII, 326 pages. 1981.

Vol. 860: Probability in Banach Spaces III. Proceedings, 1980. Edited by A. Beck. VI, 329 pages. 1981.

Vol. 861: Analytical Methods in Probability Theory. Proceedings 1980. Edited by D. Dugué, E. Lukacs, V. K. Rohatgi. X, 183 pages. 1981.

Vol. 862: Algebraic Geometry. Proceedings 1980. Edited by A. Libgober and P. Wagreich. V, 281 pages. 1981.

Vol. 863: Processus Aléatoires à Deux Indices. Proceedings, 1980. Edited by H. Korezlioglu, G. Mazziotto and J. Szpirglas. V, 274 pages. 1981.

Vol. 864: Complex Analysis and Spectral Theory. Proceedings, 1979/80. Edited by V. P. Havin and N. K. Nikol'skii. VI, 480 pages. 1981.

Vol. 865: R. W. Bruggeman, Fourier Coefficients of Automorphic Forms. III, 201 pages. 1981.

Vol. 866: J.-M. Bismut, Mécanique Aléatoire. XVI, 563 pages. 1981.

Vol. 867: Séminaire d'Algèbre Paul Dubreil et Marie-Paule Malliavin. Proceedings, 1980. Edited by M.-P. Malliavin. V, 476 pages. 1981.

Vol. 868: Surfaces Algébriques. Proceedings 1976-78. Edited by J. Giraud, L. Illusie et M. Raynaud. V, 314 pages. 1981.

Vol. 869: A. V. Zelevinsky, Representations of Finite Classical Groups. IV, 184 pages. 1981.

Vol. 870: Shape Theory and Geometric Topology. Proceedings, 1981. Edited by S. Mardešić and J. Segal. V, 265 pages. 1981.

Vol. 871: Continuous Lattices. Proceedings, 1979. Edited by B. Banaschewski and R.-E. Hoffmann. X, 413 pages. 1981.

Vol. 872: Set Theory and Model Theory. Proceedings, 1979. Edited by R. B. Jensen and A. Prestel. V, 174 pages. 1981.

Vol. 873: Constructive Mathematics, Proceedings, 1980. Edited by F. Richman. VII, 347 pages. 1981.

Vol. 874: Abelian Group Theory. Proceedings, 1981. Edited by R. Göbel and E. Walker. XXI, 447 pages. 1981.

Vol. 875: H. Zieschang, Finite Groups of Mapping Classes of Surfaces. VIII, 340 pages. 1981.

Vol. 876: J. P. Bickel, N. El Karoui and M. Yor. Ecole d'Eté de Probabilités de Saint-Flour IX - 1979. Edited by P. L. Hennequin. XI, 280 pages. 1981.

Vol. 877: J. Erven, B.-J. Falkowski, Low Order Cohomology and Applications. VI, 126 pages. 1981.

Vol. 878: Numerical Solution of Nonlinear Equations. Proceedings, 1980. Edited by E. L. Allgower, K. Glashoff, and H.-O. Peitgen. XIV, 440 pages. 1981.

Vol. 879: V. V. Sazonov, Normal Approximation - Some Recent Advances. VII, 105 pages. 1981.

Vol. 880: Non Commutative Harmonic Analysis and Lie Groups. Proceedings, 1980. Edited by J. Carmona and M. Vergne. IV, 553 pages. 1981.

Vol. 881: R. Lutz, M. Goze, Nonstandard Analysis. XIV, 261 pages. 1981.

Vol. 882: Integral Representations and Applications. Proceedings, 1980. Edited by K. Roggenkamp. XII, 479 pages. 1981.

Vol. 883: Cylindric Set Algebras. By L. Henkin, J. D. Monk, A. Tarski, H. Andréka, and I. Németi. VII, 323 pages. 1981.

Vol. 884: Combinatorial Mathematics VIII. Proceedings, 1980. Edited by K. L. McAvaney. XIII, 359 pages. 1981.

Vol. 885: Combinatorics and Graph Theory. Edited by S. B. Rao. Proceedings, 1980. VII, 500 pages. 1981.

Vol. 886: Fixed Point Theory. Proceedings, 1980. Edited by E. Fadell and G. Fournier. XII, 511 pages. 1981.

Vol. 887: F. van Oystaeyen, A. Verschoren, Non-commutative Algebraic Geometry, VI, 404 pages. 1981.

Vol. 888: Padé Approximation and its Applications. Proceedings, 1980. Edited by M. G. de Bruin and H. van Rossum. VI, 383 pages. 1981.

Vol. 889: J. Bourgain, New Classes of $\mathcal{L}^p$-Spaces. V, 143 pages. 1981.

Vol. 890: Model Theory and Arithmetic. Proceedings, 1979/80. Edited by C. Berline, K. McAloon, and J.-P. Ressayre. VI, 306 pages. 1981.

Vol. 891: Logic Symposia, Hakone, 1979, 1980. Proceedings, 1979, 1980. Edited by G. H. Müller, G. Takeuti, and T. Tugué. XI, 394 pages. 1981.

Vol. 892: H. Cajar, Billingsley Dimension in Probability Spaces. III, 106 pages. 1981.

Vol. 893: Geometries and Groups. Proceedings. Edited by M. Aigner and D. Jungnickel. X, 250 pages. 1981.

Vol. 894: Geometry Symposium. Utrecht 1980, Proceedings. Edited by E. Looijenga, D. Siersma, and F. Takens. V, 153 pages. 1981.

Vol. 895: J.A. Hillman, Alexander Ideals of Links. V, 178 pages. 1981.

Vol. 896: B. Angéniol, Familles de Cycles Algébriques - Schéma de Chow. VI, 140 pages. 1981.

Vol. 897: W. Buchholz, S. Feferman, W. Pohlers, W. Sieg, Iterated Inductive Definitions and Subsystems of Analysis: Recent Proof-Theoretical Studies. V, 383 pages. 1981.

Vol. 898: Dynamical Systems and Turbulence, Warwick, 1980. Proceedings. Edited by D. Rand and L.-S. Young. VI, 390 pages. 1981.

Vol. 899: Analytic Number Theory. Proceedings, 1980. Edited by M.I. Knopp. X, 478 pages. 1981.

Vol. 900: P. Deligne, J. S. Milne, A. Ogus, and K.-Y. Shih, Hodge Cycles, Motives, and Shimura Varieties. V, 414 pages. 1982.

Vol. 901: Séminaire Bourbaki vol. 1980/81 Exposés 561–578. III, 299 pages. 1981.

Vol. 902: F. Dumortier, P.R. Rodrigues, and R. Roussarie, Germs of Diffeomorphisms in the Plane. IV, 197 pages. 1981.

Vol. 903: Representations of Algebras. Proceedings, 1980. Edited by M. Auslander and E. Lluis. XV, 371 pages. 1981.

Vol. 904: K. Donner, Extension of Positive Operators and Korovkin Theorems. XII, 182 pages. 1982.

Vol. 905: Differential Geometric Methods in Mathematical Physics. Proceedings, 1980. Edited by H.-D. Doebner, S.J. Andersson, and H.R. Petry. VI, 309 pages. 1982.

Vol. 906: Séminaire de Théorie du Potentiel, Paris, No. 6. Proceedings. Edité par F. Hirsch et G. Mokobodzki. IV, 328 pages. 1982.

Vol. 907: P. Schenzel, Dualisierende Komplexe in der lokalen Algebra und Buchsbaum-Ringe. VII, 161 Seiten. 1982.

Vol. 908: Harmonic Analysis. Proceedings, 1981. Edited by F. Ricci and G. Weiss. V, 325 pages. 1982.

Vol. 909: Numerical Analysis. Proceedings, 1981. Edited by J.P. Hennart. VII, 247 pages. 1982.

Vol. 910: S.S. Abhyankar, Weighted Expansions for Canonical Desingularization. VII, 236 pages. 1982.

Vol. 911: O.G. Jørsboe, L. Mejlbro, The Carleson-Hunt Theorem on Fourier Series. IV, 123 pages. 1982.

Vol. 912: Numerical Analysis. Proceedings, 1981. Edited by G. A. Watson. XIII, 245 pages. 1982.

Vol. 913: O. Tammi, Extremum Problems for Bounded Univalent Functions II. VI, 168 pages. 1982.

Vol. 914: M. L. Warshauer, The Witt Group of Degree k Maps and Asymmetric Inner Product Spaces. IV, 269 pages. 1982.

Vol. 915: Categorical Aspects of Topology and Analysis. Proceedings, 1981. Edited by B. Banaschewski. XI, 385 pages. 1982.

Vol. 916: K.-U. Grusa, Zweidimensionale, interpolierende Lg-Splines und ihre Anwendungen. VIII, 238 Seiten. 1982.

Vol. 917: Brauer Groups in Ring Theory and Algebraic Geometry. Proceedings, 1981. Edited by F. van Oystaeyen and A. Verschoren. VIII, 300 pages. 1982.

Vol. 918: Z. Semadeni, Schauder Bases in Banach Spaces of Continuous Functions. V, 136 pages. 1982.

Vol. 919: Séminaire Pierre Lelong – Henri Skoda (Analyse) Années 1980/81 et Colloque de Wimereux, Mai 1981. Proceedings. Edité par P. Lelong et H. Skoda. VII, 383 pages. 1982.

Vol. 920: Séminaire de Probabilités XVI, 1980/81. Proceedings. Edité par J. Azéma et M. Yor. V, 622 pages. 1982.

Vol. 921: Séminaire de Probabilités XVI, 1980/81. Supplément Géométrie Différentielle Stochastique. Proceedings. Edité par J. Azéma et M. Yor. III, 285 pages. 1982.

Vol. 922: B. Dacorogna, Weak Continuity and Weak Lower Semicontinuity of Non-Linear Functionals. V, 120 pages. 1982.

Vol. 923: Functional Analysis in Markov Processes. Proceedings, 1981. Edited by M. Fukushima. V, 307 pages. 1982.

Vol. 924: Séminaire d'Algèbre Paul Dubreil et Marie-Paule Malliavin. Proceedings, 1981. Edité par M.-P. Malliavin. V, 461 pages. 1982.

Vol. 925: The Riemann Problem, Complete Integrability and Arithmetic Applications. Proceedings, 1979-1980. Edited by D. Chudnovsky and G. Chudnovsky. VI, 373 pages. 1982.

Vol. 926: Geometric Techniques in Gauge Theories. Proceedings, 1981. Edited by R. Martini and E.M.de Jager. IX, 219 pages. 1982.

Vol. 927: Y. Z. Flicker, The Trace Formula and Base Change for GL (3). XII, 204 pages. 1982.

Vol. 928: Probability Measures on Groups. Proceedings 1981. Edited by H. Heyer. X, 477 pages. 1982.

Vol. 929: Ecole d'Eté de Probabilités de Saint-Flour X – 1980. Proceedings, 1980. Edited by P.L. Hennequin. X, 313 pages. 1982.

Vol. 930: P. Berthelot, L. Breen, et W. Messing, Théorie de Dieudonné Cristalline II. XI, 261 pages. 1982.

Vol. 931: D.M. Arnold, Finite Rank Torsion Free Abelian Groups and Rings. VII, 191 pages. 1982.

Vol. 932: Analytic Theory of Continued Fractions. Proceedings, 1981. Edited by W.B. Jones, W.J. Thron, and H. Waadeland. VI, 240 pages. 1982.

Vol. 933: Lie Algebras and Related Topics. Proceedings, 1981. Edited by D. Winter. VI, 236 pages. 1982.

Vol. 934: M. Sakai, Quadrature Domains. IV, 133 pages. 1982.

Vol. 935: R. Sot, Simple Morphisms in Algebraic Geometry. IV, 146 pages. 1982.

Vol. 936: S.M. Khaleelulla, Counterexamples in Topological Vector Spaces. XXI, 179 pages. 1982.

Vol. 937: E. Combet, Intégrales Exponentielles. VIII, 114 pages. 1982.

Vol. 938: Number Theory. Proceedings, 1981. Edited by K. Alladi. IX, 177 pages. 1982.

Vol. 939: Martingale Theory in Harmonic Analysis and Banach Spaces. Proceedings, 1981. Edited by J.-A. Chao and W.A. Woyczyński. VIII, 225 pages. 1982.

Vol. 940: S. Shelah, Proper Forcing. XXIX, 496 pages. 1982.

Vol. 941: A. Legrand, Homotopie des Espaces de Sections. VII, 132 pages. 1982.

Vol. 942: Theory and Applications of Singular Perturbations. Proceedings, 1981. Edited by W. Eckhaus and E.M. de Jager. V, 363 pages. 1982.

Vol. 943: V. Ancona, G. Tomassini, Modifications Analytiques. IV, 120 pages. 1982.

Vol. 944: Representations of Algebras. Workshop Proceedings, 1980. Edited by M. Auslander and E. Lluis. V, 258 pages. 1982.

Vol. 945: Measure Theory. Oberwolfach 1981, Proceedings. Edited by D. Kölzow and D. Maharam-Stone. XV, 431 pages. 1982.

Vol. 946: N. Spaltenstein, Classes Unipotentes et Sous-groupes de Borel. IX, 259 pages. 1982.

Vol. 947: Algebraic Threefolds. Proceedings, 1981. Edited by A. Conte. VII, 315 pages. 1982.

Vol. 948: Functional Analysis. Proceedings, 1981. Edited by D. Butković, H. Kraljević, and S. Kurepa. X, 239 pages. 1982.

Vol. 949: Harmonic Maps. Proceedings, 1980. Edited by R.J. Knill, M. Kalka and H.C.J. Sealey. V, 158 pages. 1982.

Vol. 950: Complex Analysis. Proceedings, 1980. Edited by J. Eells. IV, 428 pages. 1982.

Vol. 951: Advances in Non-Commutative Ring Theory. Proceedings, 1981. Edited by P.J. Fleury. V, 142 pages. 1982.

Vol. 952: Combinatorial Mathematics IX. Proceedings, 1981. Edited by E. Billington, S. Oates-Williams, and A.P. Street. XI, 443 pages. 1982.

Vol. 953: Iterative Solution of Nonlinear Systems of Equations. Proceedings, 1982. Edited by R. Ansorge, Th. Meis, and W. Törnig. VII, 202 pages. 1982.